Review Questions for
NEUROANATOMY

Review Questions for
NEUROANATOMY
Structural and Functional

by

William T. Mosenthal, MD

Department of Anatomy
Dartmouth Medical School

Review Question Series
Series Editor: Thomas R. Gest, PhD
University of Arkansas for Medical Sciences

The Parthenon Publishing Group Inc.
International Publishers in Medicine, Science & Technology

One Blue Hill Plaza, Pearl River, New York 10965, USA

Published in the USA by
The Parthenon Publishing Group Inc.
One Blue Hill Plaza,
PO Box 1564, Pearl River,
New York 10965, USA

Published in Europe by
The Parthenon Publishing Group Limited
Casterton Hall, Carnforth,
Lancs LA6 2LA, UK

Copyright © 1996 The Parthenon Publishing Group

Library of Congress Cataloging-in-Publication Data

Mosenthal, William T.
　Review questions for neuroanatomy : structural and functional / by
William T. Mosenthal.
　　　p.　cm. – (Review questions series)
　ISBN: 1-85070-653-0
　1. Neuroanatomy – Examinations, questions, etc. I. Title. II. Series.
　[DNLM: 1. Neuroanatomy – examination questions. WL 18.2 M898r 1996]
　QM451.M675　1996
　611.8′076 -- DC20

　　　　　　　　　　　　　　　　　　　　　　　　　　　　96-2843
　　　　　　　　　　　　　　　　　　　　　　　　　　　　CIP

British Library Cataloguing in Publication Data

Mosenthal, William T.
　Review Questions for neuroanatomy
　1. Neuroanatomy - Examinations, questions, etc.
　I. Title II. Neuroanatomy
　611.8
　ISBN 1-85070-653-0

This edition published 1996

No part of this publication may be reproduced in any form without permission from the publishers except for the quotation of brief passages for the purpose of review.

Printed in the United States

Preface

The neuroanatomic knowledge tested in this book comprises the expertise a competent physician must obtain and use in evaluating the nervous systems of patients encountered in a general practice. No effort is made to explore the areas of current neuroanatomic research.

I wish to thank Colleen King for indefatigable secretarial help and suggestion.

Contents

1. Autonomic Nervous System (20) 1
2. Basal Ganglia (9) 6
3. Blood Supply (40) 9
4. Brain Stem (60) 20
5. Cerebellum (21) 29
6. Cortex (58) 33
7. Cranial Nerves (50) 44
8. CSF and Meninges (30) 54
9. Embryology (17) 60
10. Eye and Ear (52) 64
11. Hypothalamus (16) 75
12. Limbic System (14) 78
13. Miscellaneous (96) 81
14. Spinal Cord (74) 99
15. Thalamus (15) 114
16. Tracts (51) 117

Number in parentheses indicates number of questions available.

SECTION 1: AUTONOMIC NERVOUS SYSTEM

1.001 The pelvic splanchnic nerves ("nervi erigentes") originate from spinal nerves:
   A. L4, L5, S1.
   B. L5, S1, S2.
   C. S1, S2, S3.
   D. S2, S3, S4.
   E. S3, S4, S5.

D is the correct answer.

1.002 Regarding the innervation of the abdominal and pelvic viscera, which is INCORRECT?
   A. Both sympathetic and parasympathetic systems are importantly involved in intestinal function.
   B. Pelvic viscera receive parasympathetic innervation from pelvic splanchnic nerves contained in S2, 3, 4.
   C. The pelvic plexus contains only parasympathetic fibers.
   D. Sympathetic postganglionic fibers targeted for visceral organs arise from neurons in preaortic and pelvic ganglia.
   E. The vagus nerve is not involved in rectal and bladder function.

C is the correct answer.
The pelvic plexus is composed of pre- and postganglionic sympathetic fibers, and preganglionic parasympathetic fibers. Sympathetics which haven't synapsed in the inferior mesenteric ganglion will arrive in the pelvic plexus as preganglionics. "Sympathetic" neurons are present in the pelvic plexus for their synaptic pleasure. All the other statements are correct.

1.003 Regarding splanchnic nerves:
   A. All splanchnic nerves contain preganglionic fibers.
   B. Thoracic splanchnics arise from postganglionic neurons in the preaortic ganglia.
   C. Thoracic splanchnics synapse with postganglionic neurons located in the sympathetic trunk.
   D. All splanchnics communicate with spinal nerves by way of gray rami communicantes.
   E. Pelvic splanchnic nerves come from all sacral spinal nerves.

A is the correct answer.
All splanchnics are preganglionic. They do not bother with sympathetic trunk neurons, but pass through the trunk without synapse to reach preaortic ganglia where postganglionic neurons await. Splanchnics arise from cord segments T5-12, L1-2, and S2-3-4. They do not communicate with spinal nerves. Spinal nerves are doing business in the body periphery, splanchnics in the inner viscera.

1.004 Concerning the sympathetic system:
   A. The upper three or four thoracic segments supply sympathetic preganglionics to the head and neck. Sympathetic innervation of the heart and lungs must come from segments lower than the first four.
   B. There are four autonomic ganglia in the head containing postganglionic neurons with which incoming sympathetic preganglionics synapse.
   C. The cervical sympathetic ganglia service only the cervical spinal nerves, the blood vessels of the scalp and face, and dilator muscle of the iris.
   D. All parasympathetics concerned with autonomic functions in the head arise from the first three or four thoracic segment lateral horn cells.
   E. The vagus nerve has no parasympathetic function in the head.

E is the correct answer.
Vagal parasympathetic participation in the affairs of the head is probably nil. Parasympathetic function in the head is carried out by branches of cranial nerves III, VII, and IX (ciliary, sphenopalatine, submandibular, and otic ganglia). Lateral horn cells in the thoracic cord are not involved in this parasympathetic innervation. The sympathetics arise from the top two thoracic lateral horns, synapse and become postganglionic in the superior cervical ganglion, and project to the head riding on branches of the carotids. As there are not postganglionic sympathetic neurons in the head, obviously all sympathetics in the head are postganglionic, thanks to a visit in the superior cervical ganglion. These sympathetics play a role in controlling the blood supply to the brain and organs such as the salivary glands, as well as the functions listed in "C". The upper thoracic sympathetic outflow participates in the innervation of heart-lungs-esophagus as well as producing sympathetic innervation of head and neck structures.

1.005 Regarding the autonomic system, all the following statements are true EXCEPT:
A. Cauda equina lesions may interrupt the reflex arc that produces an automatically emptying "cord bladder".
B. The parasympathetics play a large role in temperature regulation by means of their control of peripheral vascular smooth muscle.
C. In the absence of parasympathetic inflow to the gut, difficulty of propulsion of gut content and tonic closure of the gut sphincters becomes manifest.
D. Preaortic ganglia offer synapse areas for preganglionic sympathetic fibers.
E. Sympathetic postganglionic fibers are present in every spinal nerve without exception.

1.006 Select the one best answer:
A. The stomach is innervated by the vagus alone.
B. Preganglionics to the distal small intestine are carried in pelvic splanchnic nerves.
C. Parasympathetic nerves concerned with the contraction of bladder smooth muscle are contained in the cauda equina.
D. Interruption of the spinal reflex arc that produces micturition in the newborn would occur with a transection of the spinal cord in the cervical region.
E. Sexual potency in the male requires intact internal pudendal nerves.

1.007 Concerning the autonomic system, which of the following is CORRECT?
A. Acetylcholine is the neurotransmitter between the pre- and postganglionic neurons in both the sympathetic and parasympathetic systems.
B. The transmitter substance between postganglionic sympathetic and parasympathetic terminals, and effector smooth muscle cells, is norepinephrine.
C. Postganglionic parasympathetic fibers are characteristically long as compared to postganglionic sympathetic fibers.
D. Sacral autonomic neurons synapse in the sacral portion of the sympathetic trunk, then run to the pelvic viscera as postganglionic fibers.
E. Sacral nerves have no gray rami communicantes.

B is the correct answer.
Autonomic activities peripherally are largely in the hands of the sympathetic system. Every spinal nerve has sympathetic fibers in it. Both branches of the autonomic system are at work "inside" operating the viscera. Without parasympathetic inflow the gut does not function properly; the unopposed sympathetics produce sphincter and circular muscle hypertonicity. The bladder will empty on the stimulus of distention, providing the sensory-motor reflex arc through the spinal cord is intact. The cauda equina carries parasympathetic detrusor fibers from the cord to the bladder. Preganglionic sympathetics of the thoracic and lumbar splanchnics synapse in preaortic ganglia.

C is the correct answer.
Bladder muscle contraction is stimulated by S2-3-4 parasympathetics. The S2-3-4 level of the spinal cord is well up in the lumbar vertebral region (conus medullaris is at the L1-2 level). Thus detrusor fibers from this high level must reach the bladder by traveling as fibers in the cauda equina. All the other statements are false. The stomach is served by both sympathetic and parasympathetic nerves. Pelvic splanchnics take over for the vagus in the splenic flexure region. Parasympathetics are all vagus proximal to this level and below the head. Cervical transection of the cord will not interrupt the lower reflex arc which can produce "automatic" bladder emptying. Sexual potency in the male relies on intact autonomics traveling to the penile erectile tissue from alongside the prostate, under the pubic arch, to the penis. The internal pudendal nerves are not involved in potency per se.

A is the correct answer.
Neurotransmitters at the effector sites are: acetylcholine for parasympathetics, and norepinephrine for the sympathetics. Postganglionic sympathetic fibers are the long ones; parasympathetic postganglionics are characteristically short, and thus have local effect. Sacral parasympathetic neurons characteristically synapse with postganglionic neurons located in the wall of the organ innervated. The sympathetic trunk is a sympathetic system structure. The craniosacral parasympathetics have nothing to do with it. Sacral segmental spinal nerves all receive gray rami communicantes.

1.008 Regarding the large intestine, which of the following is INCORRECT?
A. Sympathetic splanchnic nerves are inoperative if Auerbach's myenteric ganglion cells are absent.
B. All autonomic nerves reach the large intestine via colic blood vessels.
C. Parasympathetic fibers from the pelvic plexus are instrumental in proper functioning of the defecation reflex.
D. Unopposed action of the sympathetic nerves to the rectum will result in a functional obstruction of the colon due to circular muscle spasticity.
E. Lumbar truncal sympathectomy will cause widespread colonic dysfunction.

A is the correct answer.
Ganglion cells in Auerbach's plexus service incoming parasympathetic preganglionic fibers. Absence of these ganglion cells results in loss of parasympathetic function, and overactivity (unopposed) sympathetic activity. This results in functional obstruction of the colon by the involved spastic aganglionic portion of bowel (Hirschsprung's disease). Parasympathetics stimulate digestive motion in the gut (segmentation, peristalsis). This obviously includes defecation. Autonomic fibers reach the gut via the gut blood vessels. The lumbar sympathetic trunk is not vital for large intestine function.

1.009 Which of the following statements is INCORRECT? The hypothalamus may be considered the "chief ganglion" of the autonomic system. As such, it:
A. controls the functioning of the endocrine system.
B. produces appropriate visceral reactions to emotional stimuli.
C. safeguards water balance by its ability to secrete a water retaining hormone.
D. is primarily a sympathetic system center.
E. has strong connections with midbrain tegmental reticular system nuclei.

D is the correct answer.
All other statements are correct. As the Head Ganglion, the hypothalamus manages both sympathetic and parasympathetic affairs. Means of control are the pituitary and the endocrine system, and the midbrain connections with the reticular system.

1.010 Regarding the autonomic system, which is CORRECT?
A. The vagus nerve supplies parasympathetic innervation only to structures that are derived from embryonic endoderm or yolk sac.
B. The most important of the several pathways for sympathetic input to the pelvic plexus is via the hypogastric nerve/plexus.
C. The adrenal medulla is innervated by postganglionic sympathetic nerve fibers whose cell bodies are located in the preaortic ganglia.
D. Bilateral division of the pelvic splanchnic nerves may decrease acid secretion from a Meckel's diverticulum.
E. Bilateral division of the vagus commonly causes impotence.

B is the correct answer.
The GI tract is the main system derived from the yolk sac endoderm. This system does offer a significant operational challenge to the autonomic system; but how about the mesodermal heart, etc.? The adrenal medulla itself might be considered a postganglionic structure, and its secretion, adrenalin, a postganglionic neurotransmitter. Exhibition of adrenalin is stimulated by preganglionic sympathetic fibers. Sympathetic fibers reach the pelvic plexi via the hypogastric plexi. Pelvic splanchnics are S2-3-4 preganglionic parasympathetic fibers. Their loss will not affect the acid secreting mucosa of a Meckel's diverticulum - this is vagus territory. Contrariwise, the vagi neither reach nor function in the pelvis-genital areas.

1.011 Autonomic nerve fibers reach the bladder and prostate via all of the following EXCEPT:
A. pudendal nerve.
B. hypogastric nerve.
C. lumbar splanchnic nerve.
D. pelvic splanchnic nerve.
E. pelvic plexus.

A is the correct answer.
Autonomic fibers may reach the bladder-prostate in all the other structures. The pudendal nerve promptly leaves the pelvis and functions outside the pelvic bowl.

Autonomic Nervous System

1.012 A preganglionic sympathetic axon originates in the lateral horn of the gray matter of the spinal cord at the level of T3. Which statement below is INCORRECT?
- A. May synapse with a second sympathetic neuron in the sympathetic chain of ganglia.
- B. May synapse with more than one neuron in that ganglion.
- C. May synapse with a second neuron in more than one ganglion.
- D. May synapse in the cervical sympathetic ganglion chain.
- E. Will provide no visceral innervation.

E is the correct answer.
T3 preganglionic sympathetics will run to thoracic viscera after synapse in the thoracic trunk. All the other statements are true. A preganglionic sympathetic axon has many choices.

1.013 Regarding spinal nerves, select the one CORRECT statement:
- A. All spinal nerves contain fibers of the white rami communicantes.
- B. Only spinal nerves from T1 to L2 carry postganglionic sympathetic fibers.
- C. The dorsal primary ramus of T2 contains no sympathetic postganglionics because it need only innervate the deep back muscles.
- D. All spinal nerve receive gray rami from the sympathetic trunk.
- E. Parasympathetic postganglionic fibers form a significant portion.

D is the correct answer.
Sympathetic outflow comes only from T1 to L2. Thus only T1-L2 spinal nerves contain and emit white rami communicantes. However, by virtue of the sympathetic trunk, which extends from the base of the skull to the coccyx, all spinal nerves may receive post ganglionic gray rami. Parasympathetics are not included in this spinal nerve distribution. Deep back muscle blood supply and back skin sweat and oil glands require sympathetic control.

1.014 Concerning the greater splanchnic nerve, one statement below is CORRECT:
- A. It contains preganglionic parasympathetic fibers.
- B. It arises from neurons in the intermediolateral cell column (lateral horn) of the gray matter of the cord.
- C. It distributes primarily to ganglion cells in the gut tube wall.
- D. It contributes to the superior cervical ganglion.
- E. It carries autonomic fibers to the thoracic viscera.

B is the correct answer.
The greater splanchnic nerve is concerned with the innervation of abdominal viscera. It contains preganglionic fibers destined to synapse with postganglionic neurons in preaortic ganglia (not in the gut tube wall). It has no interest in cervical ganglia or thoracic viscera. Its cells of origin are indeed in the spinal cord lateral horn.

1.015 Preganglionic sympathetic fibers may be found:
- A. in the gray rami communicantes.
- B. in the musculocutaneous nerve.
- C. in the greater splanchnic nerve.
- D. on the carotid artery.
- E. in the L4 spinal nerve.

C is the correct answer.
All the other statements are incorrect in that they refer to postganglionic fibers.

1.016 The efferent innervation to the small intestine is supplied by the:
- A. dorsal motor nucleus of the vagus.
- B. nucleus ambiguus.
- C. solitary nucleus.
- D. salivatory nucleus.
- E. stellate ganglion.

A is the correct answer.
The intestine is "viscera", its innervation is GVE, supplied by the GVE dorsal motor nucleus of the vagus. Nucleus ambiguus is the vagal SVE (branchiogenic) nucleus supplying innervation to pharynx-larynx voluntary muscles. Nucleus solitarius is afferent - GVA and SVA (taste). Nucleus salivatorius is efferent - GVE to autonomics of the head. (Vagus operates below head level.) Stellate ganglion - combined inferior cervical and first thoracic sympathetic ganglia - contains sympathetic neurons with axons to the head, upper extremity, and thoracic viscera. None of these axons venture below the diaphragm.

1.017 Which of the following is most closely associated with taste?
  A. Geniculate ganglion.
  B. Sphenopalatine ganglion.
  C. Ciliary ganglion.
  D. Otic ganglion.
  E. Sympathetic trunk.

A is the correct answer.
Via the chorda tympani from the anterior two-thirds of the tongue. (The otic is not a sensory ganglion of IX.)

1.018 Which of the following is most closely associated with the cranial nerve IX?
  A. Geniculate ganglion.
  B. Sphenopalatine ganglion.
  C. Ciliary ganglion.
  D. Otic ganglion.
  E. Sympathetic trunk.

D is the correct answer.
The otic ganglion contains postganglionic neurons which, when stimulated by preganglionics in the IX nerve, power the parotid gland.

1.019 Which of the following is most closely associated with the Edinger-Westphal nucleus?
  A. Geniculate ganglion.
  B. Sphenopalatine ganglion.
  C. Ciliary ganglion.
  D. Otic ganglion.
  E. Sympathetic trunk.

C is the correct answer.
Ciliary ganglion contains postganglionics synapsing with preganglionics coming from the Edinger-Westphal nucleus.

1.020 Which of the following is most closely associated with pupillary dilation?
  A. Geniculate ganglion.
  B. Sphenopalatine ganglion.
  C. Ciliary ganglion.
  D. Otic ganglion.
  E. Sympathetic trunk.

E is the correct answer.
Pupillary dilation is a sympathetic function. Preganglionics are in T1 cord segment, postganglionics in the superior ganglion of the cervical sympathetic trunk.

SECTION 2: BASAL GANGLIA

2.001 The neostriatum:
   A. has a major functional relationship with the substantia nigra.
   B. has major interconnections with the neocerebellum through the middle cerebellar peduncle.
   C. is concerned importantly with activity of the autonomic nervous system.
   D. is constituted of two separated and non-related nuclei, the putamen and caudate.
   E. is a diencephalic structure.

A is the correct answer.
The globus pallidus (paleostriatum) and subthalamic nucleus are of diencephalic origin. The telencephalic neostriatum consists of two portions (caudate and putamen) separated for the most part by the fibers of the internal capsule, but joined rostromedially, an area to which the internal capsule does not extend. The two are the same stuff. Major connections exist between the neostriatum and the substantia nigra, and the globus pallidus, but not with the neocerebellum or the autonomic nervous system.

2.002 Regarding the corpus striatum:
   A. is necessary for normal motor function.
   B. communicates with the thalamus via the thalamic fasciculus, and therefore has significant influence on the perception and evaluation of sensory stimuli.
   C. is reciprocally related to both the substantia nigra and the subthalamic nucleus.
   D. receives extensive input from the cerebral cortex.
   E. consists of caudate, putamen, and globus pallidus.

B is the correct answer.
The corpus striatum is concerned primarily with motor, not sensory, activity. It is indeed necessary for normal motor function and depends heavily on interconnections with substantia nigra and the subthalamic nucleus to accomplish this mission. Its work orders come from the cerebral cortex; its major output is from the globus pallidus.

2.003 The lenticular fasciculus consists of fibers which:
   A. arise in the globus pallidus of the corpus striatum.
   B. cut across the internal capsule to reach the subthalamus.
   C. terminate in the VA nucleus of the thalamus.
   D. join with the fibers of the ansa lenticularis in the prerubral field to form the thalamic fasciculus.
   E. arise in the putamen of the lentiform nucleus.

E is the correct answer.
The lenticular fasciculus (along with the ansa lenticularis) is the main output of the globus pallidus. Descriptions of its course and termination in statements B, C, and D are accurate. The putamen feeds information to the globus pallidus, but the neurons producing the lenticular fasciculus reside in the globus pallidus.

2.004 Which of the following statements is false?
   A. The caudate nucleus and the putamen are termed the neostriatum.
   B. The globus pallidus and subthalamus are structurally similar. They are separated by the internal capsule through which multiple connections exist between the two.
   C. The thalamic fasciculus projects primarily to the anterior nucleus of the thalamus.
   D. The globus pallidus is called the pallidum or paleostriatum.
   E. The fasciculus lenticularis and the ansa lenticularis constitute the primary outflow of the globus pallidus.

C is the correct answer.
The thalamic fasciculus projects to the VA/VL nuclei of the thalamus. The anterior nucleus of the thalamus receives the mammillo-thalamic tract. The other statements are quite correct.

2.005 Which of the following does not terminate in the nucleus ventralis posterolateralis/medialis?
 A. Trigeminothalamic fibers.
 B. Medial lemniscus.
 C. Stria terminalis.
 D. Laryngeal sensation.
 E. Spinothalamic tract.

C is the correct answer.
The stria terminalis connects the amygdala with the hypothalamus/septal nuclei. The other items are sensory tracts relayed to consciousness by the VPL/VPM. Don't sell the vagus short. Its sensory endings in pharynx, larynx, and upper esophagus report sensory events from these areas via the laryngeal nerves to the brain stem. Here the vagus sensory fibers enter the trigeminal chief sensory nucleus and spinal tract - thence to the VPL/VPM thalamus with <u>trigeminal</u> fibers. Ever have a sore throat?!

2.006 The basal ganglia:
 A. participate in the initiation and maintenance of voluntary movements.
 B. participate in many cognitive processes.
 C. if lesioned, are associated with a resultant contralateral hemiparesis.
 D. are an extension of the reticular apparatus.
 E. are essentially vestigial structures serving very few functions in the primate nervous system.

A is the correct answer.
The basal ganglia don't help us think, don't innervate muscles; but are vitally important in the initiation and maintenance of voluntary movement. This without discernible aid from the reticular system.

2.007 Victims of Parkinsonism have a significant lack of:
 A. dopamine.
 B. acetylcholine.
 C. GABA (gamma aminobutyric acid).
 D. serotonin.
 E. noradrenalin.

A is the correct answer.
Dopamine is produced by neurons of the substantia nigra which project to neurons of the neostriatum. In the latter area, dopamine is delivered as an inhibitory neurotransmitter; controlling the activity of neostriatal neurons. Lack of dopamine results in disinhibition of neostriatal activity, resulting in the motor and tonic disorders of Parkinsonism.

2.008 Presumed normal functions of the basal ganglia are:
 A. coordination of movement and inhibition of irrelevant sensory input.
 B. maintenance of consciousness and initiation of sleep.
 C. the initiation of motor activity and automatic maintenance of ongoing well-learned motor skills.
 D. are in fact poorly defined.
 E. depend on the integrity of limbic system circuits.

C is the correct answer.
Neuronal activity can be recorded in the basal ganglia before a voluntary movement is carried out. This confirms the role of the basal ganglia in initiating voluntary movement. Without basal gangliar activity, continuation of well-learned motor skills (e.g., locomotion) is impaired. These striatal functions are well defined. Coordination of motion and inhibition of irrelevant sensory input are in the realm of cerebellar and reticular system responsibility respectively. Maintenance of consciousness, the initiation of sleep, are responsibilities of the reticular activating system. The limbic circuits operate mainly in the visceral-emotional areas.

2.009 Regarding the corpus striatum:
A. Is part of the "new" motor system characteristic of humans.
B. Dyskinesias of corpus striatal origin are thought to be the result of overactivity of structures released from inhibiting influences of a lesioned area.
C. The symptoms of paralysis agitans are due to uninhibited activity of the neostriatum secondary to a degenerative process involving dopamine-producing neurons of the substantia nigra.
D. An intact subthalamic nucleus is necessary to control output of the globus pallidus.
E. Because of crossing of the corticospinal tracts, lesions of the corpus striatum are manifest in contralateral dyskinesias.

A is the correct answer.
The corpus striatum is an "old" construction, seen in all vertebrates. The paleostriatum (globus pallidus) is older than the neostriatum (caudate-putamen), but both are older than the neocortex which integrates them into its own activity. Breaks in this integration occur with disorder of any of its parts, allowing a finely balanced system to go awry (viz. the sudden drop when you slide off your end of the see-saw). The disturbances of Parkinsonism furnish a good example of this "release" phenomenon. The subthalamic nucleus, of similar embryonic origin and still strongly connected with the globus pallidus, is essential for control of the latter's output. Subthalamic destruction results in violent "ballismus" stemming from uncontrolled globus pallidus activity. Because of ipsilateral delivery to the motor cortex and decussation of the corticospinal tract, lesions of the corpus striatum produce contralateral effects.

SECTION 3: BLOOD SUPPLY

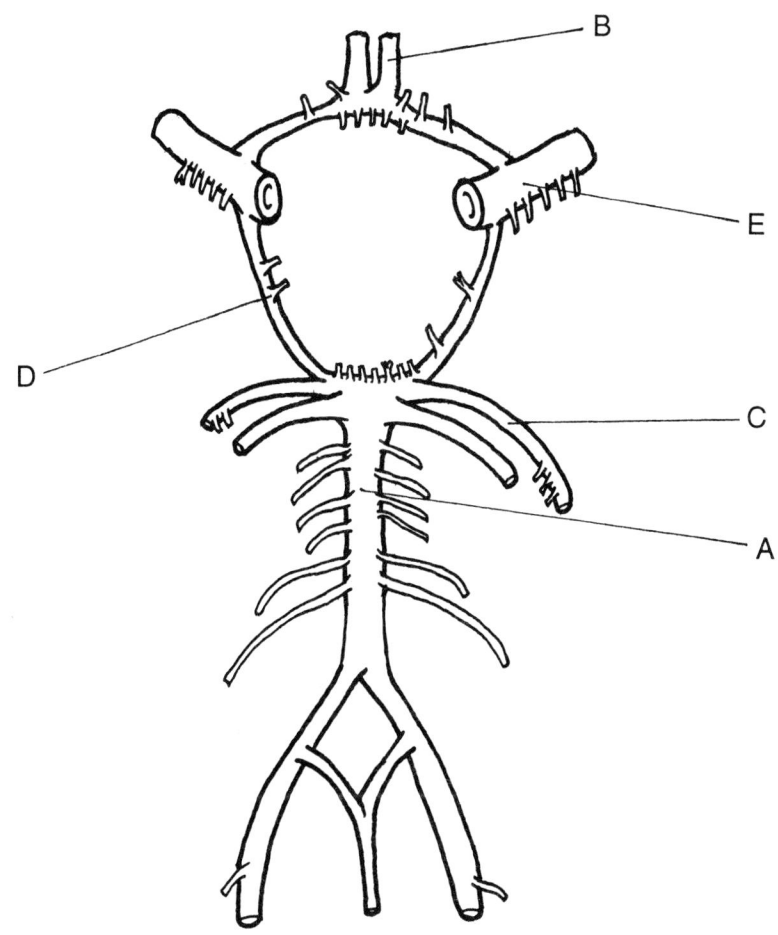

For questions 1 - 5, refer to the diagram above.

3.001 In the diagram above, ____ runs in the longitudinal cerebral fissure.

B is the correct answer.
Anterior cerebral A.

3.002 In the diagram above, ____ runs in the transverse fissure between cerebellum and stem, and cerebrum.

C is the correct answer.
Posterior cerebral A.

3.003 In the diagram above, ____ runs in lateral cerebral fissure.

E is the correct answer.
Middle cerebral A.

3.004 In the diagram above, ____ connects the posterior cerebral with the internal carotid artery.

D is the correct answer.
Posterior communicating A.

3.005 In the diagram above, ____ courses over the surface of the insula.

E is the correct answer.

10   Blood Supply

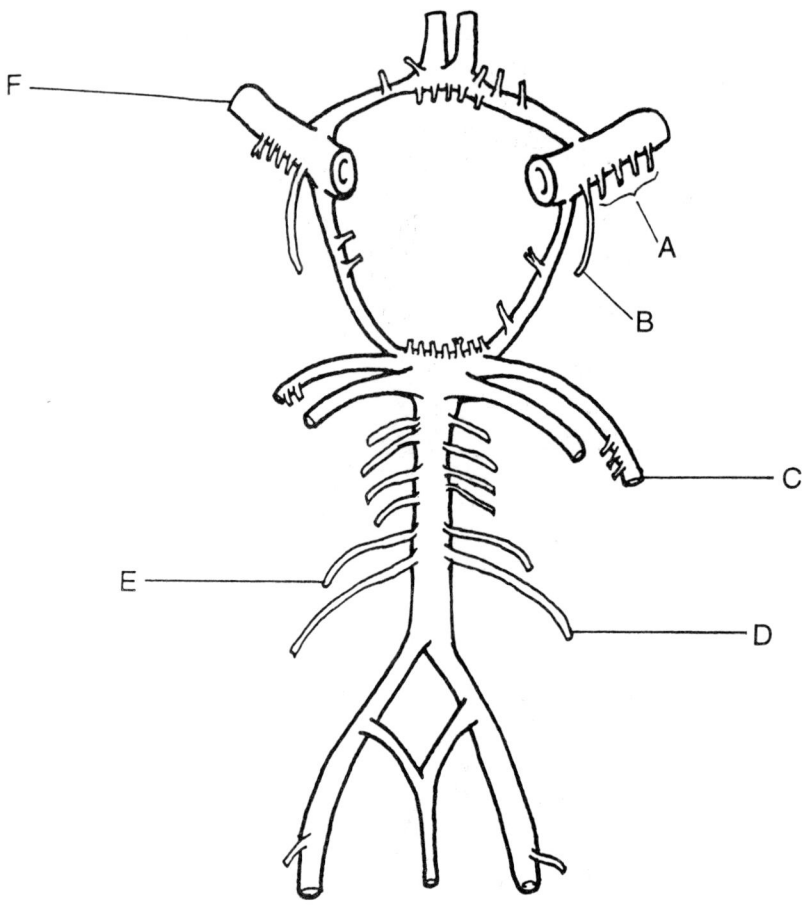

For questions 6 - 10, refer to the diagram above.

3.006 In the diagram above, which artery traverses the internal acoustic meatus and ramifies throughout the membranous labyrinth of the internal ear?

E is the correct answer.
Labrynthine A. (internal auditory A.)

3.007 In the diagram above, which artery supplies the choroid plexus of the inferior horn of the lateral ventricle?

B is the correct answer.
Anterior choroidal A.

3.008 In the diagram above, which artery, if occluded, results in a contralateral homonymous hemianopsia?

C is the correct answer.
Posterior cerebral A.

3.009 In the diagram above, which artery supplies the basal ganglia and the internal capsule?

A is the correct answer.
Lenticulostriate AA. (lateral striate AA.)

3.010 In the diagram above, which artery, if occluded, results in hemiparesis, though the lower extremity of the opposite side may not be totally paralyzed?

F is the correct answer.
Middle cerebral A.

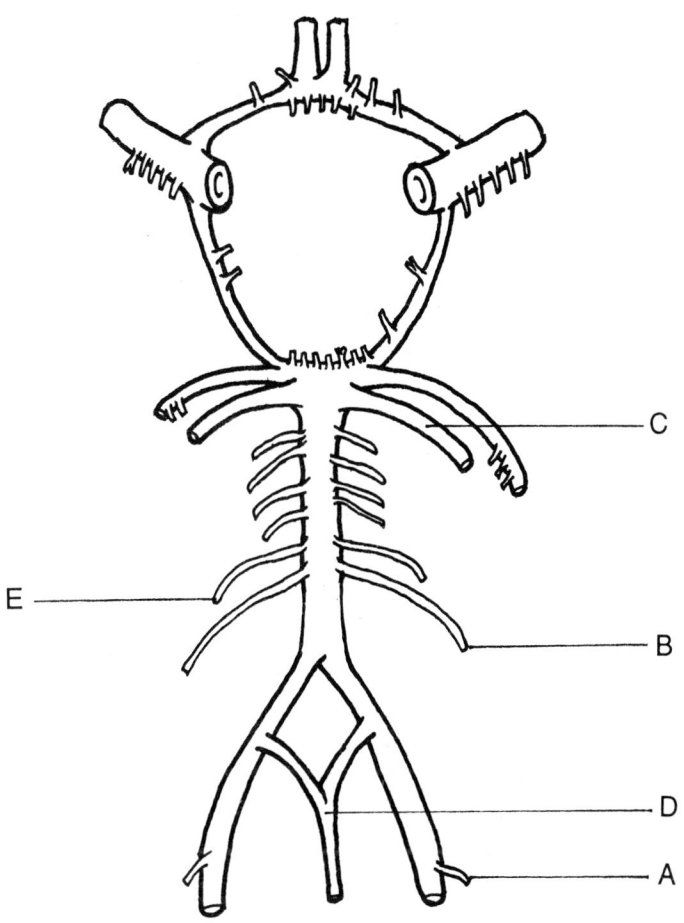

For questions 11 - 14, refer to the diagram above.

3.011 In the diagram above, which artery supplies the anterior portion of the cord.

D is the correct answer.
Anterior spinal A.

3.012 In the diagram above, which artery supplies the anterior inferior portion of the cerebellum.

B is the correct answer.
Anterior inferior cerebellar A.

3.013 In the diagram above, which artery supplies the posterior part of the cerebellar hemispheres, inferior vermis, and dorso-lateral region of the medulla oblongata.

A is the correct answer.
Posterior inferior cerebellar A.

3.014 In the diagram above, the oculomotor nerve emerges between this artery and the posterior cerebral artery.

C is the correct answer.
Superior cerebellar A.

Blood Supply

3.015 Which artery provides major blood supply to the midbrain?
    A. Anterior cerebral artery.
    B. Middle cerebral artery.
    C. Posterior cerebral artery.
    D. Anterior choroidal artery.
    E. Anterior communicating artery.

C is the correct answer.

3.016 Which of the following supplies Broca's motor speech area?
    A. Anterior cerebral artery.
    B. Middle cerebral artery.
    C. Posterior cerebral artery.
    D. Anterior choroidal artery.
    E. Anterior communicating artery.

B is the correct answer.

3.017 Which of the following supplies the visual cortex.
    A. Anterior cerebral artery.
    B. Middle cerebral artery.
    C. Posterior cerebral artery.
    D. Anterior choroidal artery.
    E. Anterior communicating artery.

C is the correct answer.

3.018 Which of the following supplies the foot area of the sensory and motor primary cortices?
    A. Anterior cerebral artery.
    B. Middle cerebral artery.
    C. Posterior cerebral artery.
    D. Anterior choroidal artery.
    E. Anterior communicating artery.

A is the correct answer.

3.019 Which of the following supplies the optic tract.
    A. Anterior cerebral artery?
    B. Middle cerebral artery.
    C. Posterior cerebral artery.
    D. Anterior choroidal artery.
    E. Anterior communicating artery.

D is the correct answer.

3.020 Which of the following supplies the medial and inferior portions of the temporal lobe?
    A. Anterior cerebral artery.
    B. Middle cerebral artery.
    C. Posterior cerebral artery.
    D. Anterior choroidal artery.
    E. Anterior communicating artery.

C is the correct answer.

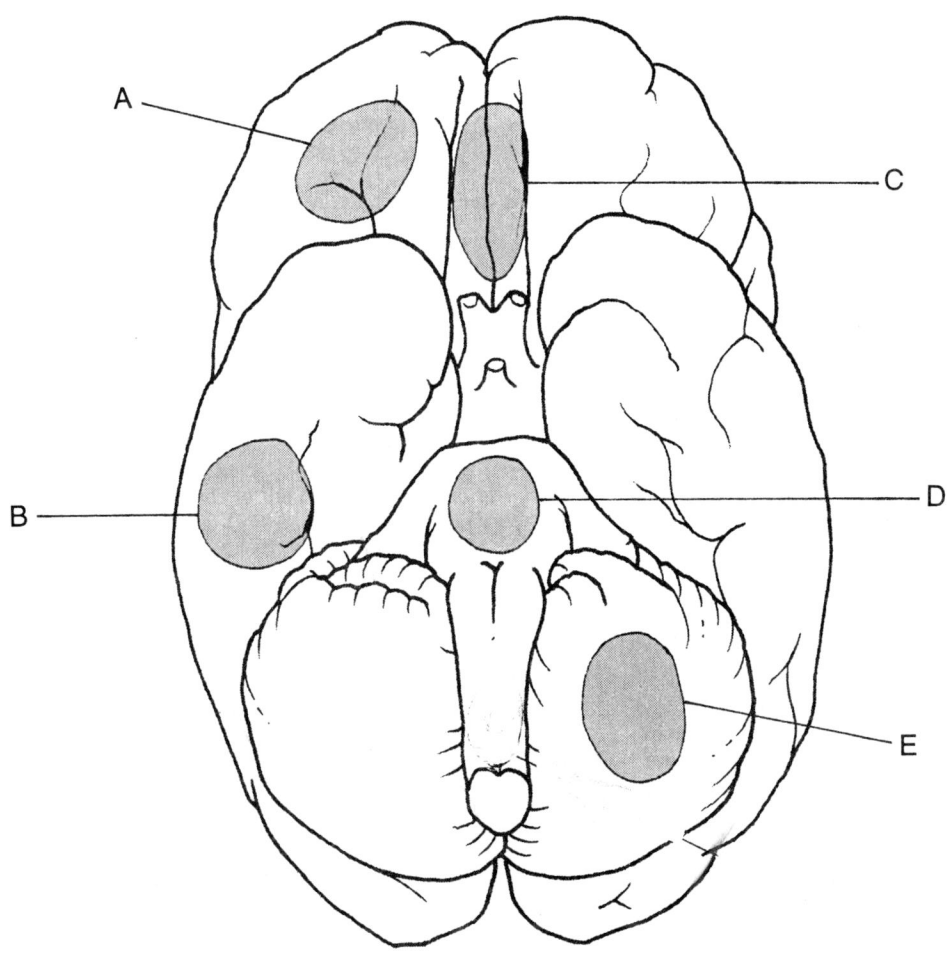

For questions 21 - 24, refer to the diagram above.

3.021   In the diagram above, ____ is middle cerebral artery territory   A is the correct answer.

3.022   In the diagram above, ____ is posterior cerebral artery territory   B is the correct answer.

3.023   In the diagram above, ____ is basilar artery territory   D is the correct answer.

3.024   In the diagram above, ____ is anterior cerebral artery territory   C is the correct answer.

14 Blood Supply

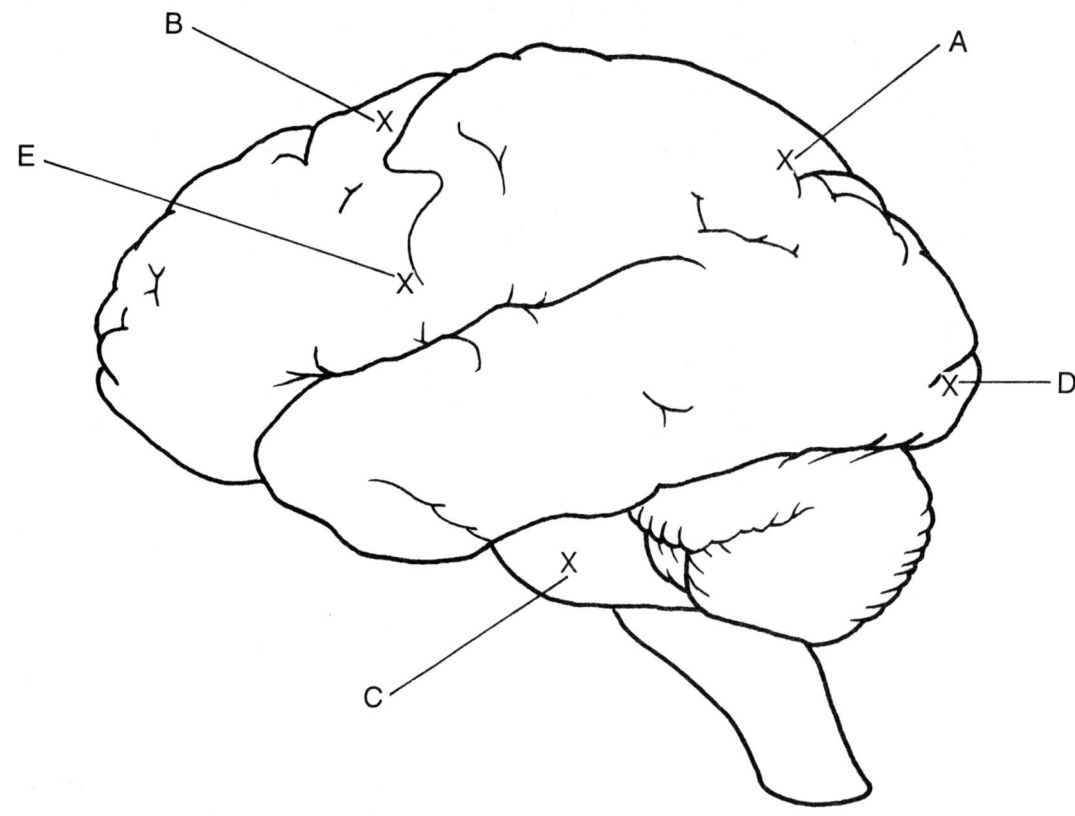

For questions 25 - 28, refer to the diagram above.

3.025 Area ____ is supplied by the basilar artery.

C is the correct answer.

3.026 Area ____ is supplied by middle cerebral artery.

E is the correct answer.
The middle cerebral actually runs in the Sylvian fissure, with its cortical branches issuing therefrom.

3.027 Area ____ is supplied by middle and anterior cerebral arteries.

B is the correct answer.

3.028 Area ____ is supplied by anterior, middle, and posterior cerebral arteries.

A is the correct answer.

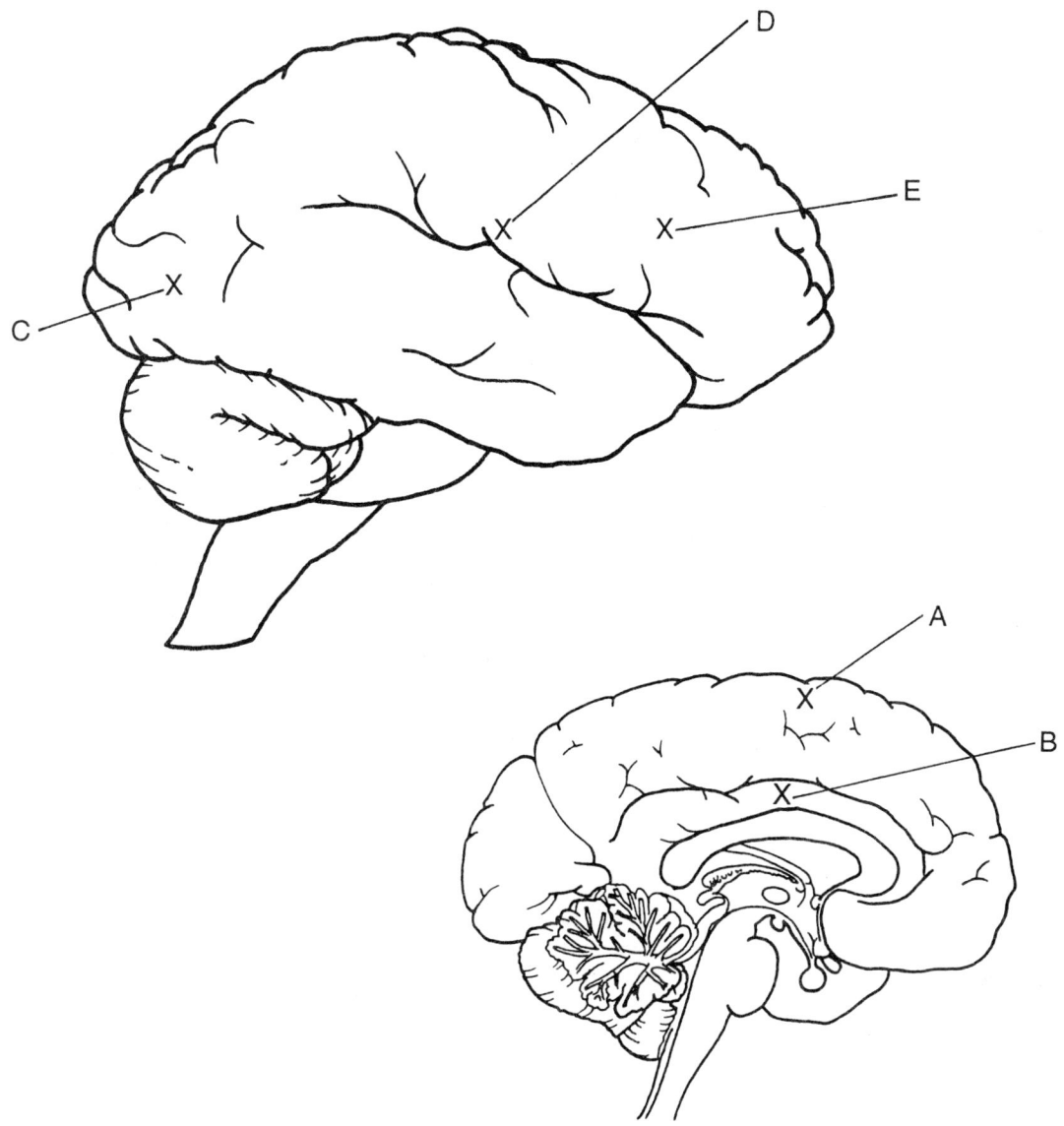

For question 29, refer to the diagram above.

3.029 With heart failure and circulatory hypotension, in which area of the cortex would the circulation be most compromised?

A is the correct answer.
With low flow, areas most ischemic will be those most distant from their arterial source. Thus area "A" is most at risk, being the farthest from both the middle and anterior cerebral arteries, which supply the area. Areas "D" and "E" are closer to the middle meningeal stem. Areas "B" and "C" are close to their supply from the anterior and posterior cerebrals respectively. The situation is reversed with occlusion of a main supplying artery, e.g., the middle meningeal. In this case, area "A" would be least likely to necrose, due to help from the anterior cerebral. Areas "D" and "E" would certainly die.

Blood Supply

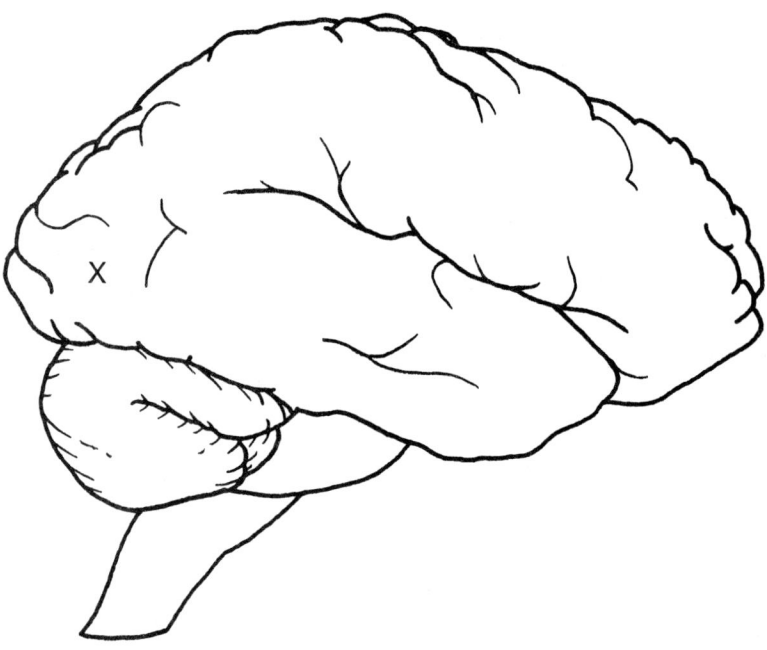

For questions 30 - 32, refer to the diagram above.

3.030 Infarct at area "X" (in diagram above) might be caused by an embolus lodging in the:
   A. middle cerebral artery.
   B. anterior cerebral artery.
   C. posterior cerebral artery.
   D. central artery of the Circle of Willis.

C is the correct answer.
"X" marks the occipital lobe, supplied by the posterior cerebral artery.

3.031 Sparing of macular vision after occlusion of one posterior cerebral artery may be due to collateral circulation provided by:
   A. anterior communicating artery.
   B. posterior communicating artery.
   C. middle cerebral artery.
   D. superior cerebral artery.
   E. contralateral posterior cerebral artery.

C is the correct answer.
Branches of the middle may keep this most important area of the calcarine cortex alive after loss of the posterior cerebral.

3.032 Which one of the following are entirely or partially supplied by the posterior cerebral artery?
   A. Precentral gyrus (Area 4) on medial surface.
   B. Choroid plexus of fourth ventricle.
   C. Postcentral gyrus (Area 3-1-2) on medial surface.
   D. Calcarine cortex.
   E. Subcallosal area.

D is the correct answer.

3.033 Select the one incorrect statement regarding the cavernous sinus.
A. Communicates with the pharyngeal and pterygoid plexi.
B. Contains valves.
C. Communicates with the superior petrosal sinus.
D. Receives blood from the ophthalmic vein.
E. May receive blood from the facial vein.

B is the correct answer.
There are no valves in any of these venous structures. Blood may flow from pharyngeal and pterygoid plexi to the facial vein and thence to the cavernous sinus or through small veins from the pterygoid sinus through the foramen ovale to the cavernous sinus. The ophthalmic vein is a direct tributary of the cavernous sinus. The superior petrosal sinus connects the cavernous with the transverse sinus. With the absence of valves, blood from any of these venous structures may flow either to or from the cavernous sinus.

3.034 Select the one incorrect statement regarding the dural sinuses.
A. Evacuate cerebrospinal fluid.
B. Communicate with scalp veins.
C. Receive blood from the deep structures of the telencephalon through the great cerebral vein of Galen.
D. Communicate with the vertebral plexus.
E. Lose all flow when the internal jugulars are occluded.

E is the correct answer.
There are adequate collateral routes for venous blood flow from the brain - especially the vertebral plexi - when both internal jugulars are occluded. Dural sinuses communicate with scalp veins through emissary veins. The great cerebral vein, carrying blood from the interior of the brain, drains into the straight sinus. Dural sinuses evacuate CSF via arachnoidal villi.

3.035 Destruction of a cavernous sinus could result in all of the following except:
A. ptosis of the eyelid.
B. anesthesia of the tongue.
C. diplopia.
D. loss of power to accommodate.
E. paralysis of the superior oblique muscle.

B is the correct answer.
Any structure within the cavernous may be involved. Thus the III, IV, and VI nerves to the extraocular muscles, and the ophthalmic and maxillary divisions of the trigeminal would be included. The mandibular branch of V would escape as it does not traverse the cavernous sinus.

3.036 A cerebral vascular lesion resulting in a contralateral hemiparesis, hemianesthesia, hemianopsia, and both motor and sensory aphasia most likely involves branches of:
A. the posterior cerebral artery.
B. the anterior cerebral artery.
C. the lenticulostriate vessels.
D. the middle cerebral artery.
E. the anterior choroidal artery.

D is the correct answer.
All of the symptoms arise in the cerebral territory of the middle cerebral artery. Posterior cerebral artery lesions at worst will not reach the anterior cerebrum with its motor strip and motor speech area. Anterior cerebral artery lesions involve the medial aspect of the hemisphere and will not produce paresis of the upper extremity and body. Lenticulostriate arteries do supply the internal capsule and could produce all the described symptoms save hemianopsia. An anterior choroidal lesion might involve the optic tract (hemianopsia) but is not widespread enough in its distribution to infarct the motor and sensory areas of one hemicortex.

3.037 A stroke caused right-sided weakness (hemiparesis) which is more marked in the arm than leg, and distally (hand-foot) greater than proximally (shoulder-trunk) indicates the site of the lesion most likely to be:
A. in the middle cerebral artery distribution of the left cerebral cortex.
B. in the right lateral corticospinal tract.
C. in the left medullary pyramid.
D. in the left cerebral peduncle.
E. anywhere in the descending corticospinal system.

A is the correct answer.
The distinguishing feature here is the fact that hemiparesis is more marked in the arm than in the leg. To produce this difference the lesion would have to be in the cerebral hemisphere where the anterior cerebral artery collateral supply to the middle cerebral can keep the leg going to a degree. The lesion must be in the left middle cerebral artery distribution to produce right-sided symptoms. Lesions in the left cerebral peduncle, the left medullary pyramid, or anywhere in the left descending corticospinal system would produce right hemiparesis without favoring the leg.

18 Blood Supply

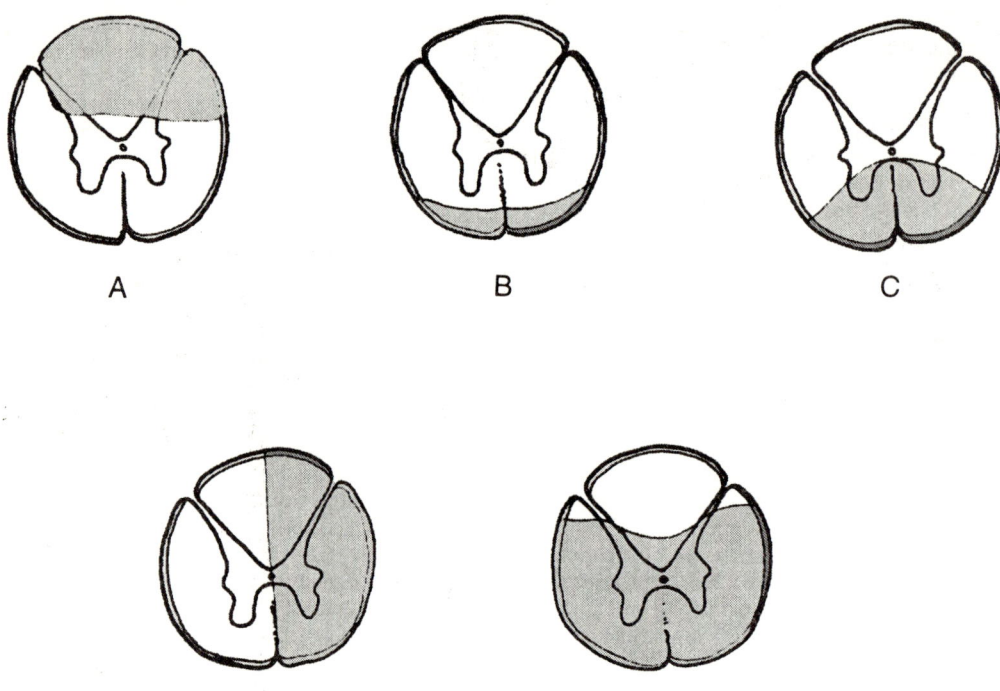

For question 38, refer to the diagram above.

3.038 Due to a severe compression fracture of the lower thoracic spine, the anterior spinal artery was injured and thrombosed. Which of the diagrams above most accurately shows the likely ischemic destruction of the spinal cord.

E is the correct answer.
The anterior spinal artery is responsible for all the area shown destroyed in "E". There is little help from the posterior spinals. Obviously a devastating injury!

3.039 An ischemic infarction in the distribution of the left middle cerebral artery may cause all of the following deficits EXCEPT:
  A. pinprick and temperature loss over the left trunk and limbs.
  B. Broca's motor aphasia.
  C. right homonymous hemianopia.
  D. Wernicke's receptive aphasia.
  E. loss of stereognosis in the right hand.

A is the correct answer.
Housed in the middle cerebral artery distribution are the primary motor and sensory strips (pre- and postcentral gyri) from knee to head, influencing the right side of the body; Broca's and Wernicke's areas (almost always in the left hemisphere), and the right visual field radiations. There is no middle cerebral artery influence over the left bodily sensory or motor functions.

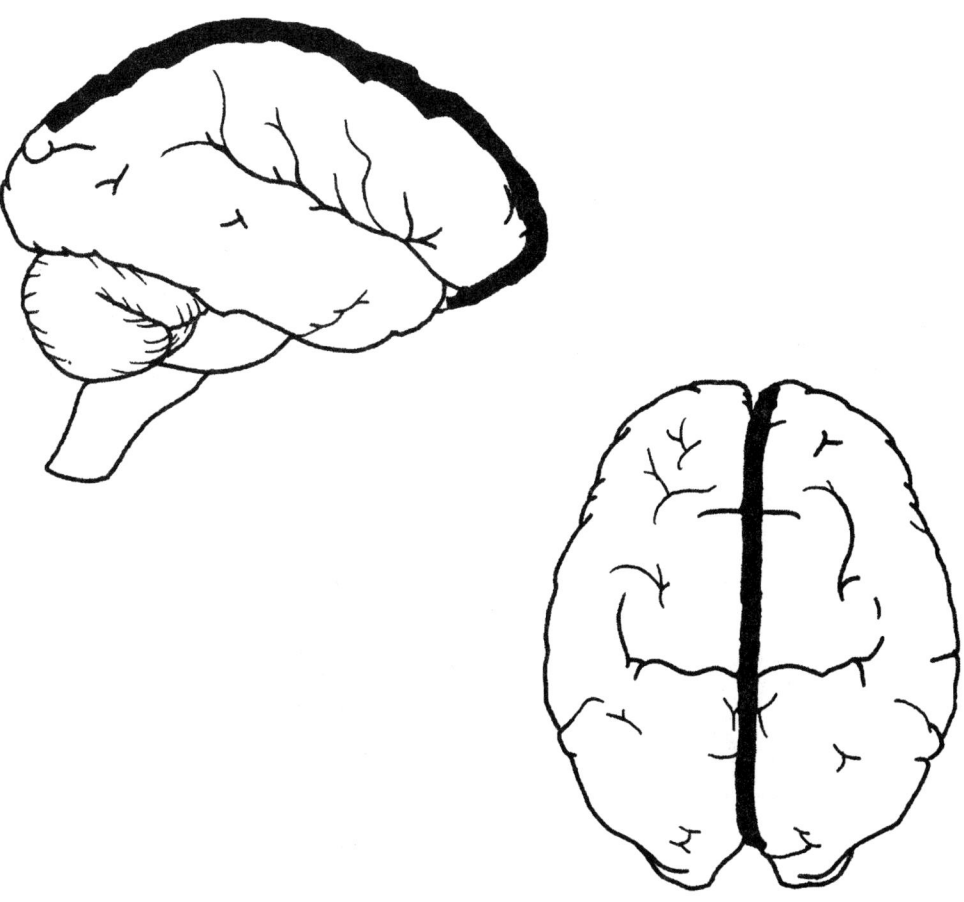

For question 40, refer to the diagram above.

3.040 The infarction (blackened area) demonstrated in the superior view of the brain above is in the distribution of the:
    A. anterior cerebral artery.
    B. middle cerebral artery.
    C. posterior inferior cerebral artery.
    D. posterior cerebral artery.
    E. posterior inferior cerebellar artery.

A is the correct answer.
The anterior cerebral artery area of supply includes the medial aspect of the hemisphere, and extends slightly up over the junction of the medial and lateral hemispheric aspects.

SECTION 4: BRAIN STEM

4.001 Select the discordant pair:
A. Betz cells --- corticobulbar tract.
B. IX nerve --- gag reflex.
C. Brachium conjunctivum --- VL thalamus.
D. Visceral sensation --- nucleus solitarius.
E. MLF --- medial lemniscus.

E is the correct answer.
The medial longitudinal fasciculus contains fibers from the vestibular and reticular systems coordinating head and eye movements and conjugate gaze. The medial lemniscus conveys proprioceptive and stereognostic sensation to the VPL thalamus and thence to the sensory cortex. The two are in no way concordant!

4.002 The facial nucleus:
A. receives cortical innervation from corticopontine fibers via the brachium pontis.
B. is situated in the pontine tegmentum.
C. innervates musculature derived from branchial arches.
D. has centers controlling upper facial and lower facial musculature.
E. is in the same cell column as the nucleus ambiguus and the trigeminal motor nucleus.

A is the correct answer.
Everything is correct save for one word in A. The innervation of bulbar motor nuclei is by corticobulbar fibers. (The brachium pontis is not involved.)

4.003 The decussation of the brachium conjunctivum is located in the:
A. rostral diencephalon.
B. caudal midbrain.
C. rostral pons.
D. caudal pons.
E. medulla.

B is the correct answer.
By the time the brachium conjunctivum reaches the rostral midbrain and the red nucleus, its decussation has been completed.

4.004 Which of the following structures is found in the tectum of the midbrain?
A. Superior colliculus.
B. Medial longitudinal fasciculus.
C. Fibers of the oculomotor nerve.
D. Ventral (anterior) spinocerebellar tract.
E. All of the above.

A is the correct answer.
All but the superior colliculus are in the "core", the tegmentum of the midbrain.

4.005 Which of the following statements is CORRECT regarding the cerebral aqueduct?
A. Contains the choroid plexus.
B. Connects the fourth ventricle and spinal canal.
C. If obstructed, produces internal hydrocephalus.
D. Is a remnant of the rhombencephalon.
E. Is part of the subthalamus.

C is the correct answer.
The cerebral aqueduct contains no choroid plexus, connects the third and fourth ventricles, is a relic of the mesencephalic vesicle, and is not part of the diencephalic subthalamus.

4.006 Which of the following structures is not located in the tegmentum of the pons?
A. Superior olivary nucleus.
B. Inferior olivary nucleus.
C. Spinal tract and nucleus of the trigeminal nerve.
D. MLF.
E. The reticular nucleus - PPRF.

B is the correct answer.
Inferior olive is in the medulla. All the others are pontine.

4.007 Concerning the superior colliculus, which statement is INCORRECT?
A. Receives input from the retina.
B. Is in the tectum of the midbrain.
C. Plays a role in the coordination of head-eye movements.
D. Contains the third cranial nerve nucleus.
E. Is nourished by blood from the posterior cerebral artery.

D is the correct answer.
The superior colliculus is part of the midbrain tectum. The oculomotor nucleus is down in the tegmentum of the midbrain. All the other statements are correct.

4.008 Select the discordant pair:
A. MLF --- medial vestibulospinal tract.
B. Corticobulbar --- precentral gyrus.
C. Brachium conjunctivum --- dentate nucleus.
D. Nucleus solitarius --- "hunger pains".
E. Flocculus --- corticopontine fibers.

E is the correct answer.
Flocculus is archicerebellum. It deals with the vestibular apparatus and head motion. The corticopontine tract is a "new" development communicating with the "new" cerebellum to help plan and perform "new" sophisticated motions of the extremities. The medial vestibulospinal fibers course throughout the cervical spine in the MLF to reflexly coordinate head position and the central visual axis. Corticobulbars issue from the precentral gyrus; brachium conjunctivum in great part from the dentate nucleus. The nucleus solitarius accepts all such general visceral afferent stimuli.

4.009 Which of the following statements is not located in the tegmentum of the pons?
A. Trapezoid body.
B. Corticospinal tract.
C. Spinothalamic tract.
D. Tectospinal tract.
E. MLF.

B is the correct answer.
The tegmentum of the pons does not welcome such newcomers as the corticospinal tracts and the brachium pontis. All the other structures belong to the pontine tegmentum.

4.010 The medial lemniscus originates from the:
A. dorsal cochlear nucleus.
B. nucleus gracilis.
C. ventral cochlear nucleus.
D. superior olivary nucleus.
E. trapezoid body.

B is the correct answer.
The medial lemniscus carries conscious proprioceptive information from the spinal cord posterior columns. The gracile and cuneate nuclei transmit this information to the VPL thalamus via the medial lemniscus.

4.011 Pick the one CORRECT statement:
A. Fibers of the fasciculus cuneatus have cell bodies in dorsal root ganglia of T6 and down caudally; and on the contralateral side.
B. Fibers of the ventral corticospinal tract cross at the junctional area of the medulla and spinal cord.
C. The spinothalamic tract has cell bodies in contralateral dorsal root ganglia.
D. The lateral corticospinal tract is a "recent" evolutionary development, and is responsible especially for fine movements of the distal extremities.
E. The ventral corticospinal tract is not concerned with gamma loop function.

D is the correct answer.
Cuneatus is T6 and up. Fibers of the ventral corticospinal tract are those pyramidal fibers that do not cross here. Spinothalamic tract cell bodies are in the contralateral substantial gelatinosa. All "upper" motor neuron tracts influence the gamma loop.

4.012 Select the discordant pair:
A. Nucleus dorsalis (Clarke) --- spinocerebellum.
B. Medial lemniscus --- internal arcuate fibers.
C. Unconscious proprioception --- ventral posteromedial nucleus of thalamus.
D. Tractus solitarius --- visceral sensation.
E. Hypoglossal nerve --- general somatic efferent.

C is the correct answer.
The VPM serves to communicate and relay sensory stimuli to the cerebral cortex and consciousness. The other statements are all compatible. The tongue muscles are derived from somites, not branchial arches. Their nerve supply is GSE.

4.013 Which of the following is located immediately beneath the floor of the fourth ventricle?
  A. Abducens nucleus.
  B. Motor nucleus of the facial nerve.
  C. Spinothalamic tract.
  D. Medial lemniscus.
  E. Nucleus ambiguus.

A is the correct answer.
The facial nerve lies just under the floor as it arches over the abducens nucleus. The two produce the facial colliculus. The SVA facial and ambiguus nuclei are forced down into the tegmentum by the GVE and GSE nuclei present in the immediate subventricular position - dorsal motor of vagus and abducens. Spinothalamic tract is always found in the anterolateral tegmentum, well below the ventricle. The medial lemniscus occupies the middle area of the tegmentum at this level.

4.014 Which of the following will lie lateral to the sulcus limitans (alar plate derivation)?
  A. Hypoglossal.
  B. Trochlear.
  C. Facial.
  D. Chief sensory nucleus of trigeminal.
  E. Dorsal motor nucleus vagus.

D is the correct answer.
Alar plate produces sensory elements. All but D are motor.

4.015 The decussation of the medial lemniscus (internal arcuate fibers) is located in the:
  A. spinal cord.
  B. medulla.
  C. pons.
  D. midbrain.
  E. diencephalon.

B is the correct answer.
Vital anatomic wisdom!

4.016 In the medulla, fibers of the hypoglossal nerve:
  A. course between the pyramid and the inferior olivary nucleus.
  B. decussate to the opposite side.
  C. course between the restiform body and the spinal nucleus of the fifth cranial nerve.
  D. course between the inferior olivary nucleus and the cuneate tubercle.
  E. emerge at the dorsolateral sulcus.

A is the correct answer.
The close relation of the hypoglossal nerve and the pyramid affords vital information in understanding a crossed paralysis involving ipsilateral tongue and contralateral body.

4.017 Which of the following is NOT part of the diencephalon?
  A. Mammillary body.
  B. Subthalamic nucleus.
  C. VPL nucleus.
  D. Uncus.
  E. Habenula.

D is the correct answer.
The uncus is part of the telencephalic parahippocampal gyrus.

4.018 Which of the following nuclei form parts of the GVE cell column?
  A. Superior olivary nucleus.
  B. Nucleus solitarius.
  C. Dorsal motor nucleus of the vagus.
  D. Nucleus ambiguus.
  E. Nucleus of the trigeminal spinal tract.

C is the correct answer.
Items A, B, and E are sensory nuclei. Nucleus ambiguus is the nucleus of the branchogenic pharynx-larynx musculature supplied by the vagus (SVE).

4.019 Which of the following statements regarding the nucleus of the spinal tract of the trigeminal is CORRECT?
- A. It is confined to the medulla.
- B. It is so-called because it sends fibers to the spinal cord nuclei.
- C. Axons of the spinal trigeminal nucleus are distributed to the thalamus as fibers in the spinal tract of the trigeminal.
- D. It has characteristics shared with the dorsal gray horn of the spinal cord, especially with the substantia gelatinosa.
- E. It receives stretch receptors from the muscles of mastication.

D is the correct answer.
The trigeminal spinal tract nucleus extends from the pons down to the upper cervical cord where it merges with the substantia gelatinosa of the cervical dorsal horn. The two form the long cell column registering pain and temperature stimuli for the entire body. Axons from this column cross and ascend to the VPL/VPM nuclei of the thalamus as the spinothalamic and trigeminothalamic tracts. Stretch receptors from the muscles of mastication are handled in the mesencephalic root nucleus of the trigeminal.

4.020 The junction of the midbrain and thalamus is represented by a plane passing through which of the following?
- A. Point just caudal to the inferior colliculus.
- B. Posterior commissure.
- C. Rostralmost fibers of the pons.
- D. Infundibulum.
- E. Point just cranial to the mammillary bodies.

B is the correct answer.

4.021 In the caudal medulla at the level of the sensory decussation one would see:
- A. gracile and cuneate nuclei.
- B. brachium conjunctivum.
- C. superior olivary nucleus.
- D. transverse pontine fibers.
- E. obex of the fourth ventricle.

A is the correct answer.
This section is well caudal to the fourth ventricle, the inferior olive, and the pontine brachium conjunctivum and the pontine fibers.

4.022 At the level of the inferior olive, one would see all but one of the following.
- A. Nucleus of the spinal tract of trigeminal.
- B. Nucleus ambiguus.
- C. Hypoglossal nucleus.
- D. Vertically arranged medial lemniscus.
- E. Ventral corticospinal tract.

E is the correct answer.
The ventral corticospinal tract is formed as the corticospinal pyramid decussates at the medullo-spinal junction. All the others are bona fide residents in the upper medulla. The hypoglossal nucleus produces the hypoglossal trigone in the caudal fourth ventricle. It does extend a bit down the medulla caudal to the fourth ventricle.

4.023 Which of the following nuclei is seen in the medulla?
- A. Trochlear.
- B. Superior olive.
- C. Facial.
- D. Abducens.
- E. Ambiguus.

E is the correct answer.
Trochlear nucleus is in the midbrain. Facial, abducens, and superior olive are all pontine nuclei.

4.024 All of the following would be seen in the rostral pons EXCEPT:
- A. Motor nucleus of trigeminal.
- B. Dorsal motor nucleus of vagus.
- C. Chief sensory nucleus of trigeminal.
- D. Brachium conjunctivum.
- E. Pontine nuclei.

B is the correct answer.
The dorsal motor nucleus of the vagus produces the vagal trigone in the caudal medullary floor of the fourth ventricle.

4.025 One of the following would not be seen in a section through the inferior colliculus.
- A. Trochlear nerve nucleus.
- B. Decussation of the brachium conjunctivum.
- C. Oculomotor nerve nucleus.
- D. Lateral lemniscus.
- E. Brachium of the inferior colliculus.

C is the correct answer.
This level neatly houses the auditory input and output of the inferior colliculus. The trochlear nerve nucleus is always found deep to this colliculus. The large decussation of the brachium conjunctivum is a hallmark of the proximal midbrain.

4.026 Which statement is INCORRECT regarding the red nucleus?
- A. Receives input from the cerebellum.
- B. Is part of an "old" motor system.
- C. Is connected to anterior horn cells by the reticular system.
- D. Activity can be seen after destruction of the posterior limb of the internal capsule.
- E. Communicates with the cerebellum via the inferior olive.

C is the correct answer.
The red nucleus has its own rubrospinal tract with connections to anterior horn cells throughout the spinal cord. An "old" motor nucleus, it has strong connections with the inferior olive, and receives fibers from "old" parts of the cerebellum. Loss of cortical inhibitory integration allows "old" motor efforts of the red nucleus to appear - flexor rigidity.

4.027 The tractus solitarius contains proximal processes of primary afferent nerve fibers from cell bodies in ganglia of which of the following?
- A. III.
- B. VII.
- C. XI.
- D. C2.
- E. I.

B is the correct answer.
The solitarius receives GVA stimuli from cranial nerves and SVA taste from VII, (IX, and X)..

4.028 Neurons in which of the following nuclei may be considered to constitute a "final common path" or "lower motor neuron"?
- A. Nucleus ambiguus.
- B. Inferior olivary nucleus.
- C. Dorsal motor nucleus vagus.
- D. Solitary nucleus.
- E. Edinger Westphal nucleus.

A is the correct answer.
Only ambiguus actually sends fibers to (branchogenic) muscle fibers. Autonomic preganglionic fibers do not quality, as the "effector" fiber comes from peripheral ganglia. Solitarius is sensory. The inferior olive, whatever it does, is not directly connected to muscle or gland.

4.029 The nucleus ambiguus:
- A. is found in the pons.
- B. innervates muscles derived from cervical somites.
- C. gives rise to portions of three cranial nerves - IX, X, XI (cranial portion).
- D. forms a bulge in the medial eminence of the fourth ventricle floor.
- E. contains preganglionic neurons of the vagus.

C is the correct answer.
Ambiguus is the nucleus of origin of fibers running to branchogenic musculature, is SVE. It is deep in the medullary tegmentum and innervates muscles of the third (IX), fourth and sixth (X and cranial XI) branchial arches. It contains no autonomic system neurons.

4.030 Sensory fibers of the trigeminal nerve are found in all of the following areas EXCEPT:
- A. medulla.
- B. pons.
- C. midbrain.
- D. uppermost cord segments.
- E. thalamus.

E is the correct answer.
The long trigeminal nerve has sensory fibers from the upper cervical cord to the midbrain (mesencephalic root). Secondary fibers of course do reach the thalamus.

4.031 The substantia nigra is most closely associated with:
- A. Globus pallidus
- B. Vestibular nuclei
- C. Dentate nucleus
- D. Anterior horn
- E. Dopamine

E is the correct answer.

4.032 The neocerebellum is most closely associated with:
- A. Globus pallidus
- B. Vestibular nuclei
- C. Dentate nucleus
- D. Anterior horn
- E. Dopamine

C is the correct answer.

4.033 The precentral gyrus is most closely associated with:
A. Globus pallidus
B. Vestibular nuclei
C. Dentate nucleus
D. Anterior horn
E. Dopamine

D is the correct answer.

4.034 The gamma loop is most closely associated with:
A. Globus pallidus
B. Vestibular nuclei
C. Dentate nucleus
D. Anterior horn
E. Dopamine

D is the correct answer.

4.035 The archicerebellum is most closely associated with:
A. Globus pallidus
B. Vestibular nuclei
C. Dentate nucleus
D. Anterior horn
E. Dopamine

B is the correct answer.

4.036 The brachium of the superior colliculus is most closely associated with:
A. Fornix columns
B. Prerubral field
C. Pretectal nucleus
D. Ventral posteromedial nucleus thalamus

C is the correct answer.

4.037 The thalamic fasciculus is most closely associated with:
A. Fornix columns
B. Prerubral field
C. Pretectal nucleus
D. Ventral posteromedial nucleus thalamus

B is the correct answer.

4.038 The mammillothalamic tract is most closely associated with:
A. Fornix columns
B. Prerubral field
C. Pretectal nucleus
D. Ventral posteromedial nucleus thalamus

A is the correct answer.

4.039 The brachium conjunctivum is most closely associated with:
A. Fornix columns
B. Prerubral field
C. Pretectal nucleus
D. Ventral posteromedial nucleus thalamus

B is the correct answer.

4.040 Toothache is most closely associated with:
A. Fornix columns
B. Prerubral field
C. Pretectal nucleus
D. Ventral posteromedial nucleus thalamus

D is the correct answer.

4.041 The ventral spinocerebellar tract is most closely associated with:
A. Superior medullary velum
B. Nucleus ambiguus
C. Nucleus gracilis
D. Vagal trigone
E. Abducens nucleus

A is the correct answer.

4.042 The posterior columns are most closely associated with:
   A. Superior medullary velum
   B. Nucleus ambiguus
   C. Nucleus gracilis
   D. Vagal trigone
   E. Abducens nucleus

C is the correct answer.

4.043 Heart rate is most closely associated with:
   A. Superior medullary velum
   B. Nucleus ambiguus
   C. Nucleus gracilis
   D. Vagal trigone
   E. Abducens nucleus

D is the correct answer.

4.044 Vocal cord paralysis is most closely associated with:
   A. Superior medullary velum
   B. Nucleus ambiguus
   C. Nucleus gracilis
   D. Vagal trigone
   E. Abducens nucleus

B is the correct answer.

4.045 The facial colliculus is most closely associated with:
   A. Superior medullary velum
   B. Nucleus ambiguus
   C. Nucleus gracilis
   D. Vagal trigone
   E. Abducens nucleus

E is the correct answer.

4.046 The paleocerebellar cortex is most closely associated with:
   A. Flocculonodular lobe
   B. Nucleus interpositus (nuc. globose & emboliform)
   C. Fasciculus thalamicus
   D. Dentate nucleus
   E. Globus pallidus

B is the correct answer.

4.047 Efferents from striatum are most closely associated with:
   A. Flocculonodular lobe
   B. Nucleus interpositus (nuc. globose & emboliform)
   C. Fasciculus thalamicus
   D. Dentate nucleus
   E. Globus pallidus

E is the correct answer.

4.048 Nucleus fastigii is most closely associated with:
   A. Flocculonodular lobe
   B. Nucleus interpositus (nuc. globose & emboliform)
   C. Fasciculus thalamicus
   D. Dentate nucleus
   E. Globus pallidus

A is the correct answer.

4.049 Ansa and fasciculus lenticularis together form the:
   A. Flocculonodular lobe
   B. Nucleus interpositus (nuc. globose & emboliform)
   C. Fasciculus thalamicus
   D. Dentate nucleus
   E. Globus pallidus

C is the correct answer.

4.050 Nucleus ventralis lateralis is most closely associated with:
   A. Flocculonodular lobe
   B. Nucleus interpositus (nuc. globose & emboliform)
   C. Fasciculus thalamicus
   D. Dentate nucleus
   E. Globus pallidus

D is the correct answer.

4.051 Which of the following projects to nucleus ventralis anterior
   A. Flocculonodular lobe
   B. Nucleus interpositus (nuc. globose & emboliform)
   C. Fasciculus thalamicus
   D. Dentate nucleus
   E. Globus pallidus

C is the correct answer.

4.052 A stroke involves one nucleus ambiguus. Associated findings will be:
   A. Difficulty swallowing, hoarseness, and weakness of elevation of the ipsilateral shoulder and turning of the head contralaterally.
   B. Hypotension and bradycardia.
   C. Difficulty swallowing, hoarseness, and hypomotility of the bowel.
   D. Difficulty swallowing, hoarseness, and decreased taste over the posterior third of the tongue.
   E. Difficulty swallowing and hoarseness.

E is the correct answer.
The nucleus ambiguus axons are motor to the branchogenic musculature of the larynx and pharynx. Sensory - e.g. taste - functions are not involved. Bowel function is autonomic system powered, with fibers emanating from the dorsal motor nucleus of the vagus, and lateral horn neurons of the spinal cord. The same may be said for cardiovascular function. Trapezius and sternocleidomastoid muscles are branchogenic, but have their own spinal accessory nerve.

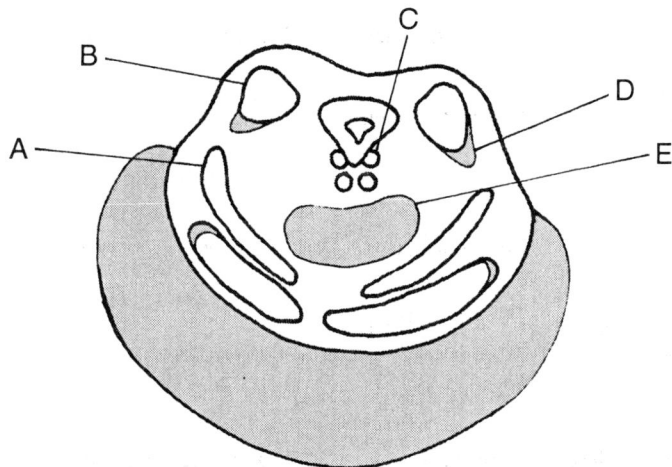

For questions 53 - 55, refer to the diagram above.

4.053 In the diagram above, _____ carries stereognostic information to the thalamus.

A is the correct answer.

4.054 In the diagram above, _____ carries cerebellar output to the thalamus.

E is the correct answer.

4.055 In the diagram above, _____ is a motor nucleus.

C is the correct answer.
The section is midbrain, level of the inferior colliculus. A. is the medial lemniscus, B. is the inferior colliculus, C. is the trochlear nucleus, D. is the lateral lemniscus, E. is the decussating brachium conjunctivum.

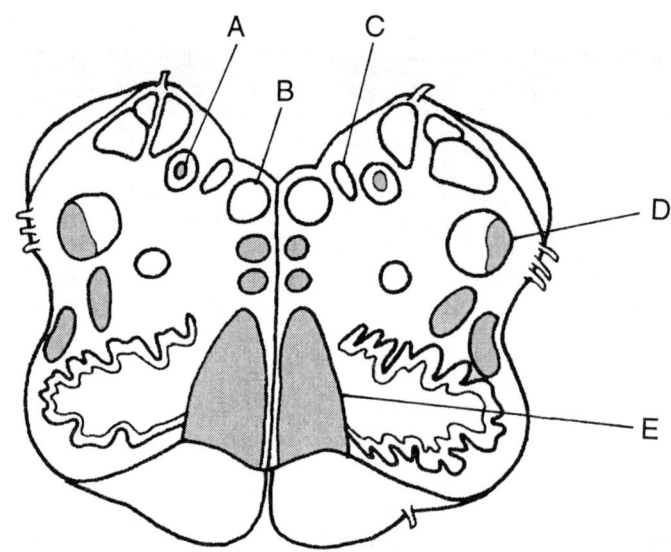

For questions 56 - 58, refer to the diagram above.

4.056 In the diagram above, _____ receives stimuli from the carotid body.

4.057 In the diagram above, _____ controls the nervous phase of gastric acid secretion.

4.058 In the diagram above, _____ contains pain fibers from the ipsilateral side of the body.

A is the correct answer.

C is the correct answer.

D is the correct answer.
Section is through the upper medulla. A. is the nucleus and tractus solitarius, B. is the hypoglossal nucleus, C. is the dorsal motor nucleus of the vagus, D. is the nucleus and spinal tract of the trigeminal, E. is the crossed medial lemniscus.

4.059 Which pairing is CORRECT?
  A. Nucleus ambiguus - SVE.
  B. Solitary nucleus - GSA.
  C. Trochlear nucleus - SVE.
  D. Facial motor nucleus - GSE.
  E. None of the above.

A is the correct answer.
The correct pairings are: A. Nucleus ambiguus - SVE, B. Solitary nucleus - GVA, C. Trochlear nucleus - GSE, D. Facial motor nucleus - SVE.

4.060 What brainstem nucleus receives all visceral afferents carried by the cranial nerves?
  A. Solitary nucleus.
  B. Nucleus ambiguus.
  C. Dorsal motor nucleus of the vagus.
  D. Salivatory nucleus.
  E. Chief sensory nucleus of the trigeminal.

A is the correct answer.
B, C, and D are motor nuclei: SVE, GVE, GVE respectively. The chief sensory nucleus of the trigeminal is concerned with touch and conscious proprioception from the face, GSA.

SECTION 5: CEREBELLUM

5.001 When the association cortex calls upon the cerebellum for a motor pattern, which of the following tract segments are involved? One statement is INCORRECT.
  A. Corticopontine fibers.
  B. Pontocerebellar fibers - via transverse fibers of the pons and brachium pontis.
  C. The cerebellar cortex, via granular cells and their parallel fibers rising to the molecular layer where they stimulate the dendrites of the Purkinje cells.
  D. Purkinje cell axons synapsing in the VL of the thalamus.
  E. Dentate nucleus.

D is the correct answer.
Purkinje axons synapse with the cerebellar roof nuclei - they do not leave the cerebellum. The roof nuclear axons voyage to the ventrolateral nucleus of the thalamus. All the other statements are correct.

5.002 Regarding the superior cerebellar peduncle, select the INCORRECT statement:
  A. Connects the cerebellum and the thalamus.
  B. Constitutes almost the entire output of the cerebellum.
  C. Its fiber content projects in greatest part to the VL thalamus.
  D. Contains fibers from muscle spindles, tendon organs, and skin exteroceptors.
  E. Contains fibers from the contralateral inferior olive.

E is the correct answer.
Inferior olivary fibers, of uncertain function in the human, supply the bulk of the fiber content of the inferior peduncle. The superior peduncle does give passage to afferents from the ventral spinocerebellar tract from muscle spindles, tendon organs, and skin exteroceptors as practically its sole afferent content. The superior peduncle is the essential outflow tract of the cerebellum, terminating largely and most importantly in the ventrolateral (VL) nucleus of the thalamus.

5.003 The anterior vermis in humans:
  A. is predominantly an archicerebellar zone.
  B. represents the axial parts of the body (i.e., trunk, neck, head).
  C. has only neocerebellar components.
  D. receives projection predominantly from the vestibular apparatus.
  E. contains neurons projecting to the dentate nucleus.

B is the correct answer.
The anterior vermis is considered part of the paleocerebellum (spinocerebellum). Its afferents are largely from muscle spindles, tendon organs, and skin pressure receptors. Its influence is largely exerted upon axial musculature maintaining extensor muscle tone, equilibrium, and posture.

5.004 The left brachium pontis is made up of fibers from neurons in the:
  A. left cerebral cortex.
  B. right cerebral cortex.
  C. thalamus.
  D. right pontine nuclei.
  E. left cerebellar hemisphere.

D is the correct answer.
Right corticopontine nuclear axons cross in the left brachium pontis to reach the left neocerebellum. From the cerebral cortex, corticopontine tracts terminate in the pontine nuclei, not in the cerebellum proper. The thalamus does not send afferents to the cerebellum. The cerebellar cortex exits via the ipsilateral dentate nucleus and the superior cerebellar peduncle.

5.005 Regarding the archicerebellum, one statement is INCORRECT:
  A. is also known as the vestibulocerebellum.
  B. consists of the flocculus and the connected vermal nodule.
  C. entering fibers relay primarily to the dentate nucleus.
  D. is concerned primarily with equilibrium.
  E. is phylogenetically the oldest part of the cerebellum.

C is the correct answer.
From the evolutionary aspect, the archicerebellum developed and functioned long before the dentate nucleus appeared. The other statements accurately describe the archicerebellar situation.

5.006 With disorders of the cerebellum:
   A. Tremor at rest is prominent.
   B. Ability to perform rapid rhythmic alternating movements is unimpaired.
   C. Due to decussation of the brachium conjunctivum, symptoms are most marked on the side opposite the site of disorder.
   D. Patient is unable to walk in a straight line.
   E. Strength is impaired.

D is the correct answer.
Motor activity is monitored and coordinated by the intact cerebellum. With cerebellar disorders, ataxia is prominent, including the inability to walk a straight line, inability to perform "difficult" motor actions such as rapid rhythmic alternating movements, and inability to perform smooth continuous purposeful movements without overshoot and undershoot (intention tremor). At rest, without motor activity, there is no tremor. Muscle strength per se is not impaired with cerebellar lesions. The brachium conjunctivum does decussate, but so does the outflow from the cortex. Thus activity is returned to the side of the cerebellar disorder. Left cerebellum to right thalamus and cortex; right cortex to left side muscles.

5.007 The greatest input to the cerebellum arises from:
   A. reticular formation.
   B. vestibulospinal tract.
   C. superior cerebellar peduncle.
   D. pontine nuclei.
   E. corticospinal tracts.

D is the correct answer.
By far the greatest input is to the greatest part of the cerebellum, the neocerebellar hemispheres. These are fed by the pontine nuclei.

5.008 The inferior cerebellar peduncles contain which of the following? (There is one INCORRECT.)
   A. Dorsal (posterior) spinocerebellar tract.
   B. Cerebellar-vestibular fibers (fastigiobulbar).
   C. MLF.
   D. Olivocerebellar fibers.
   E. Vestibulocerebellar fibers.

C is the correct answer.
Only the MLF is not an inferior cerebellar peduncle feeder. Cerebellovestibular fibers are the lone output present in the inferior peduncle. These contain archicerebellar instructions to the vestibular nuclei and reticular tracts of the cord.

5.009 The inferior cerebellar peduncle contain which of the following?
   A. Dorsal (posterior) spinocerebellar tract.
   B. Rubrospinal fibers.
   C. Corticobulbar fibers.
   D. The brachium pontis.
   E. Corticopontine fibers.

A is the correct answer.
The dorsal spinocerebellar tract furnishes important stimuli to the cerebellum from muscle spindles and tendon organs via the inferior peduncle. Rubrospinal fibers are bound for anterior horn cells, corticobulbar fibers for cranial nerve nuclei - not the cerebellum. Corticopontine fibers synapse with pontine nuclei and enter the cerebellum through the middle peduncle, the brachium pontis.

5.010 Regarding the cerebellum, one of the following is INCORRECT:
   A. An impulse coming into the cerebellum via a mossy fiber will synapse on a granular cell.
   B. Climbing fibers terminate on Purkinje cells.
   C. The cell bodies of climbing fibers are located in the inferior olivary nuclei.
   D. Fibers from Purkinje cells form the brachium conjunctivum.
   E. The brachium conjunctivum terminates on either the red nucleus or the nucleus ventralis lateralis.

D is the correct answer.
Purkinje cell axons project to cerebellar roof nuclei. The brachium conjunctivum is the outflow from the roof nuclei to the red nucleus and VL of the thalamus. Climbing fibers by definition arise in the inferior olives and terminate on Purkinje cell bodies and dendrites. Mossy fibers terminate on granule cells.

5.011 A five-year old child is brought to your clinic because of the recent onset of unsteadiness. On examination, you note nystagmus and ataxia of gait. An MRI (magnetic resonance imaging) might show:
  A. frontal lobe tumor.
  B. compression of the cervical spinal cord by a tumor.
  C. a tumor rising in the region of the flocculonodular lobe.
  D. tumor of the parietal lobe.
  E. aneurysm of the Circle of Willis.

C is the correct answer.
These symptoms are classically those of vestibular and cerebellar dysfunction (archicerebellum). A cerebellar-pontine angle tumor is a likely suspect. In this area the flocculus and the vestibular nerve (VIII) are central inhabitants. The other suggestions don't fill the symptomatic bill.

5.012 The dentate nucleus:
  A. projects to the contralateral VL (ventrolateral) nucleus of the thalamus via the brachium conjunctivum.
  B. is a paleocerebellar structure which relays Purkinje cell activity upstream via the superior cerebellar peduncle.
  C. receives inhibitory input from muscle spindles.
  D. is present in all vertebrates.
  E. is a neocerebellar structure which relays Purkinje cell activity upstream via the middle cerebellar peduncle.

A is the correct answer.
The brachium conjunctivum contains the outflow tract of the neocerebellum. As the nucleus of the neocerebellum, the dentate nucleus neurons supply its neocerebellar fiber content. Superior peduncular fibers cross in the midbrain and terminate in the ventrolateral nucleus of the thalamus. A much smaller portion of the superior peduncle is supplied by the paleocerebellar nuclei (interpositus and fastigii). These fibers end in the red nucleus. Purkinje cell output ends in the roof nuclei. The dentate, as the nucleus of a new evolutionary development engendered by the development of the sophisticated movements of the higher vertebrates, is not necessary and does not appear in the primitive vertebrates.

5.013 The cerebellum is involved in all but one of the following:
  A. planning and control of movement.
  B. maintenance of equilibrium.
  C. coordination of muscle action.
  D. awareness of pain and temperature sensations.
  E. postural reflexes.

D is the correct answer.
The cerebellum is primarily a vital cog in the activity of the motor system. It does not concern itself directly with pain and temperature stimuli.

5.014 Corticopontine tracts are most closely associated with:
  A. Inferior cerebellar peduncle
  B. Brachium conjunctivum
  C. Brachium pontis
  D. Flocculonodular lobe
  E. Vermis

C is the correct answer.
Corticopontine tracts terminate in pontine nuclei in the brachium pontis.

5.015 Fastigial nucleus is most closely associated with:
  A. Inferior cerebellar peduncle
  B. Brachium conjunctivum
  C. Brachium pontis
  D. Flocculonodular lobe
  E. Vermis

E is the correct answer.
Fastigii developed pari-passu with the development of the vermis.

5.016 Archicerebellum is most closely associated with:
  A. Inferior cerebellar peduncle
  B. Brachium conjunctivum
  C. Brachium pontis
  D. Flocculonodular lobe
  E. Vermis

D is the correct answer.
The most primitive part of the cerebellum; responding to the lateral line balancing sensory organ of fish.

5.017 Dentate nucleus is most closely associated with:
A. Inferior cerebellar peduncle
B. Brachium conjunctivum
C. Brachium pontis
D. Flocculonodular lobe
E. Vermis

B is the correct answer.
Dentate nucleus axons exit via the brachium conjunctivum.

5.018 Vestibular semicircular canals are most closely associated with:
A. Inferior cerebellar peduncle
B. Brachium conjunctivum
C. Brachium pontis
D. Flocculonodular lobe
E. Vermis

D is the correct answer.
The human representation of the piscine lateral line balance organ.

5.019 Most of the cerebellar output is in the form of:
A. Purkinje axons.
B. brachium pontis efferents.
C. mossy fibers.
D. axons originating in cerebellar nuclei.
E. internal arcuate fibers.

D is the correct answer.
Purkinje axons do not leave the cerebellum; rather, they synapse in the cerebellar roof nuclei. Cerebellar output consists of axons from these nuclear structures. Mossy fibers and brachium pontis efferents enter the cerebellum. The internal arcuates are decussating axons from gracile and cuneate nuclei of the medulla.

5.020 Select the most CORRECT association:
A. Pontine nuclei - inferior cerebellar peduncle.
B. Flocculonodular lobe - red nucleus.
C. Dorsal spinocerebellar tract - superior cerebellar peduncle.
D. Inferior olive - climbing fibers.
E. Dentate nucleus - archicerebellum.

D is the correct answer.
The large cerebellar input from the inferior olives - "climbing fibers" - (of uncertain function in the human) reaches the cerebellum via the inferior cerebellar peduncle. Pontine nuclear axons enter the cerebellum via the middle peduncle. The flocculonodular lobe reports to the fastigial nucleus, and has no direct association with the red nucleus. The dorsal spinocerebellar tract enters the cerebellum in the inferior peduncle. The dentate nucleus is the crowning glory of the neocerebellum.

5.021 Symptoms of unilateral cerebellar lesions include:
A. ipsilateral loss of position sense.
B. contralateral hyperreflexia.
C. ipsilateral paralysis with only rare chance of recovery.
D. ipsilateral intention tremor.
E. hemiballism.

D is the correct answer.
The cerebellum is concerned with ipsilateral motor coordination. Cerebellar lesions produce neither paralysis nor sensory disturbance. Hemiballism is seen with subthalamic lesions.

SECTION 6: CORTEX

6.001 Pick the discordant pair:
A. Calcarine sulcus - visual cortex.
B. Insula - primary auditory cortex.
C. Postcentral gyrus - primary somesthetic cortex.
D. Uncus of parahippocampal gyrus - primary olfactory area.
E. Precentral gyrus - motor cortex.

B is the correct answer.
The insular cortex is of uncertain functional import. It certainly is close to the auditory cortex anatomically, but is not involved in any known way with audition. The other pairings are concordant.

6.002 Movement is mediated through the primary motor cortex.
A. No other cerebral processes are necessary.
B. The hypothalamus and red nuclei play an important part in the initiation of movement.
C. The basal ganglia and cerebellum are brought into play prior to the recording of activity in the primary motor cortex.
D. Activity is recorded in the claustrum before any changes occur in the motor cortex.
E. The mammillary bodies, as part of the limbic system, play a key role in motor cortex control.

C is the correct answer.
Recording of neuronal electric activity is first seen in the so-called supplementary motor area in the medial frontal cortex, before a movement is carried out. Similarly the basal ganglia and cerebellum are found active before a movement is actually stimulated by firing of neurons of the primary motor strip. None of these other structures offered (hypothalamus, red nuclei, claustrum, mammillary bodies) play a significant role in the initiation of movement or control of the motor cortex. The claustrum is of uncertain function. It is most likely associated with sensory activity.

6.003 Destruction of the left calcarine cortex and the splenium of the corpus callosum:
A. Causes agraphia (inability to write) without alexia (inability to comprehend written words).
B. Causes a left homonymous hemianopia and alexia without agraphia.
C. Causes a right homonymous hemianopia and alexia without agraphia.
D. Causes only alexia without agraphia.
E. Causes right homonymous hemianopia and aphasia.

C is the correct answer.
This injury obviously produces a right homonymous hemianopsia. But in addition, a visual alexia (word blindness) occurs. Visual stimuli must come from the right calcarine cortex when the left visual side is knocked out. Right calcarine visual stimuli reach the left-sided Wernicke's area via fibers in the corpus callosal splenium. With destruction of the splenium, there is no way these right-sided visual recordings can reach the left-sided Wernicke's area. Thus words seen cannot be comprehended - comprehension demands activity in an intact left side speech center. Writing is not disturbed. This can be done "blindfolded".

6.004 Consciousness (awareness of self and environment):
A. Is a function of both the reticular system and the cerebral hemispheres.
B. Is the domain of the reticular formation alone.
C. Can be maintained in the absence of a functioning reticular system.
D. Can be maintained in the absence of the cerebral cortex.
E. Is a mental function without clear-cut anatomical substrates.

A is the correct answer.
Consciousness demands an intact reticular activating system plus a functioning cortex. Minus either of these --- .

6.005 Regarding the receptive speech center, one of the following is INCORRECT:
A. Is centered about the angular and supramarginal gyri.
B. Is critical for the successful understanding of the written word.
C. Is almost always located in the left cortical hemisphere.
D. Its function is not required for reading.
E. Is connected to the motor speech area by the arcuate portion of the superior longitudinal fasciculus.

D is the correct answer.
The receptive speech center, in its angular and supramarginal gyral locus, abuts visual, auditory, and somatosensory association areas. Making sense of stimuli received from these areas is the job of the receptive speech center. Communication with the motor speech center is via the arcuate fasciculus. Except for one or two percent of left-handed humans, the receptive and motor speech centers are located in the left cerebral cortical hemisphere.

6.006 A lesion in the Wernicke's area::
A. eradicates the ability to swear.
B. causes difficulty expressing words.
C. involves injury in the frontal lobe.
D. is associated with difficulty thinking in words.
E. causes loss of recent memory.

D is the correct answer.
Wernicke's area is the receptive speech area (angular and supramarginal gyri and the caudal portion of the superior temporal gyrus). With injury here, the ability to comprehend words is lost. Broca's area in the frontal lobe (opercular and triangular areas of the inferior temporal gyrus) is left "on its own", and becomes quite loquacious. There is no problem releasing words, but they make no sense - fluent jargon. "Meaningful" swearing is impossible. Recent memory is a job for the hippocampi.

6.007 A lesion in Broca's area:
A. causes difficulty in verbal language comprehension.
B. causes difficulty in verbal language execution (expression).
C. is associated with alexia.
D. is associated with a right homonymous hemianopsia.
E. is often seen with occlusive disease of the posterior cerebral artery.

B is the correct answer.
An intact Broca's motor speech center is required for the expression of coherent verbal language. Comprehension resides in Wernicke's receptive speech center. Alexia, the inability to comprehend the written word (word blindness), does not occur with an intact receptive center. Visual field defects may be seen with Wernicke's area lesions as the optic radiations are very close.

6.008 Regarding the cerebral cortex, which of the following statements is INCORRECT?
A. The association cortex of the human brain is far more extensive than all the primary sensory and motor cortices combined.
B. Association cortices are anatomically situated adjacent to their related primary cortices.
C. Broca's speech area (areas 44-45 Brodmann) occupies the opercular and triangular portions of the inferior frontal gyrus.
D. The angular gyrus forms part of Wernicke's area.
E. The auditory association area lies in the hippocampal formation.

E is the correct answer.
The auditory association area lies in the superior temporal gyrus (areas 42-22 Brodmann). All the other statements are correct.

6.009 The best data support that voluntary left-side movement is initiated:
- A. From the primary motor cortex (area 4) on the right.
- B. From a combination of basal ganglia on the right, and the left cerebellum.
- C. From the association cortices of the right hemisphere.
- D. From the appropriate segmental apparati (anterior horn cells and cranial nerve nuclei) on the left.
- E. From all the above.

C is the correct answer.
The key word is "initiated". The "idea" for a voluntary movement comes from the association cortices to the so-called supplementary motor area (SMA). This is situated in the medial area of the frontal lobe just above the cingulate gyrus, and just anterior to the "leg" portion of the primary motor cortex (area 4). In this case of a left-sided movement, the right SMA is involved. Here is where the first record of neuronal electrical activity occurs when a movement is considered. The "idea" then becomes form and action through SMA stimulation of basal ganglia and cerebellum; thence the primary motor strip; and finally the anterior horn cells or cranial nerve nuclei. The described right and left-sidedness in all statements is correct.

6.010 Intracranial structures that follow the curve of the lateral ventricle includes all but one of the following:
- A. Hypothalamic sulcus.
- B. Fornix.
- C. Caudate nucleus.
- D. Stria terminalis.
- E. Choroid plexus.

A is the correct answer.
The hypothalamic sulcus extends in linear fashion across the medial aspect of the thalamus, from the area of the posterior commissure to the area of the foramen of Monro. All other structures named describe the "big C" prescribed by the curvature of the lateral ventricle.

6.011 The right cerebral hemisphere in most adults:
- A. is an important substrate for imparting emotionality to speech.
- B. is predominant for grammatical function.
- C. is more involved in rational than emotional problems.
- D. communicates visual impressions to the left primary visual cortex.
- E. will be the source of almost complete recovery of verbal language when the left (dominant) hemisphere is destroyed.

A is the correct answer.
Recovery of verbal language after left hemisphere destruction is only possible in very young children, and those rare left-handers with bilateral speech centers. Ordinarily, the right cerebral hemisphere contributes emotionality and musicality to speech. Communication between primary visual cortices is not direct. The right primary visual cortex communicates its visual impressions to the receptive speech center via the corpus callosum.

6.012 Destruction of the amygdala on both sides in wild animals:
- A. causes loss of vocalization.
- B. always results in a lowered threshold for rage reactions.
- C. may cause a taming effect.
- D. has little effect on behavior.
- E. results in loss of temperature control.

C is the correct answer.
The cortical amygdala may be considered the "supervisor" or regulator of the diencephalic hypothalamus. Bilateral destruction of the amygdala may indeed have a taming effect, as well as the production of sexual behavior changes, feeding changes, etc. No effect on temperature control mechanisms has been demonstrated. Rage reactions are usually diminished. No effect on vocalization is seen.

6.013 Bilateral hippocampal lesions:
- A. cause hypersexuality.
- B. cause depression of consciousness.
- C. are associated with an inability to imprint new information.
- D. cause a major loss of past memory.
- E. cause a decrease in pituitary function.

C is the correct answer.
The major deficit following bilateral hippocampal lesions is the loss of recent memory. New information cannot be retained. Normally, after imprinting new information, the new material is soon stored in appropriate cortical association areas. This stored memory is retained even after subsequent hippocampal destruction. No effect on consciousness, sexual function, or pituitary function is to be expected.

6.014 The olfactory bulb of one side:
  A. distributes its fibers to the opposite cortex.
  B. distributes to both olfactory cortices (because of its connections through the corpus callosum).
  C. causes loss of smell, if destroyed.
  D. is an extradural structure.
  E. distributes to both olfactory cortices (because of its connections through the anterior commissure).

E is the correct answer.
The olfactory bulbs and tracts of each side are connected by fibers in the anterior commissure. Thus smell stimuli from each side reach both ipsilateral and contralateral olfactory cortices. If one bulb is destroyed, there will be no noticeable loss of smell (unless each side is tested individually). The corpus callosum is not involved.

6.015 Concerning the preoptic nuclear area, all the following are true EXCEPT:
  A. Receives input from the hippocampus via the fornix.
  B. Is connected to the midbrain by the medial forebrain bundle.
  C. Contributes to the stria medullaris thalami.
  D. Is situated in the posterior part of the hypothalamus.
  E. Is connected with the amygdala by the stria terminalis.

D is the correct answer.
The preoptic area is located where hypothalamus and cortex merge - thus in the anterior part of the hypothalamus. All the other statements are correct.

6.016 The motor cortex (area 4 - precentral gyrus):
  A. is arranged somatotopically with the cranial (face) representation lying in close proximity to the leg representation.
  B. is the origin of close to 100 percent of the fibers of the corticospinal tracts.
  C. is arranged somatotopically with the face representation lying near the lateral (Sylvian) Fissure and the foot representation lying close to the cingulate gyrus.
  D. has an inconsistently arranged somatotopic arrangement.
  E. when stimulated, will typically cause pain and fragmented twitches of individual muscles on the ipsilateral side of the body.

C is the correct answer.
The somatotopic arrangement of the precentral gyrus is consistently set up as in answer "C". When stimulated, it will produce muscle twitches on the contralateral side, and no pain. (These studies were done under local anesthesia down through the dura.) The corticospinal tracts actually contain fibers from the premotor area (area 6) and from the parietal lobe (areas 3-1-2, 5, and 7). About 50 percent of the corticospinal fibers arise from the precentral gyrus - area 4.

6.017 Which of the following is not a primary motor or sensory area?
  A. Brodmann area 4.
  B. Brodmann area 3, 1, 2.
  C. Brodmann area 17.
  D. Brodmann area 6.
  E. Brodmann area 41, 42.

D is the correct answer.
Brodmann 4 is the primary motor strip - precentral gyrus. Brodmann 3-1-2 is the primary somatosensory strip - postcentral gyrus. Brodmann 17 is the primary visual cortex - Calcarine cortex adjacent to the calcarine sulcus - cuneus and lingual gyri. Brodmann 6 is the premotor area, not a primary motor area. It occupies the caudal areas of the superior, middle, and inferior frontal gyri. Brodmann 41-42 is the primary auditory area - the transverse temporal gyri of Heschl buried in the floor of the Sylvian fissure superior to the superior temporal gyrus.

6.018 Pick the discordant pair:
  A. Amygdala - auditory system.
  B. Calcarine sulcus - visual cortex.
  C. Postcentral gyrus - primary somesthetic cortex.
  D. Primary auditory cortex - temporal lobe.
  E. Cingulate gyrus - limbic system.

A is the correct answer.
No anatomic connection here!

6.019 Which of the following statements regarding the ascending reticular activating system (RAS) is FALSE?
   A. Ascending fibers of the central group of reticular nuclei of the brainstem are part of the system.
   B. The ascending RAS is not functionally connected with the sensory association area of the parietal lobe.
   C. Stimulation of central reticular nuclei brings about awakening in the sleeping individual.
   D. Lesions of the system produce coma.
   E. Lesions of the ascending RAS will block recognition of somatic and auditory sensory stimuli, even though these sensory pathways are intact.

B is the correct answer.
The ascending reticular activating system is an alerting system which obviously must play upon cortical sensory areas. Without it, there is unconsciousness. Its origin is in the reticular nuclei of the brainstem – the central portion.

6.020 Broca's area is most closely associated with:
   A. Posterior cerebral artery.
   B. Seat of intelligence.
   C. Comprehension.
   D. Reading.
   E. Expression.

E is the correct answer.
The expressive speech center.

6.021 The angular gyrus is most closely associated with:
   A. Posterior cerebral artery.
   B. Seat of intelligence.
   C. Comprehension.
   D. Reading.
   E. Expression.

D is the correct answer.
Visual recordings of written words are funneled to this area for understanding and comprehension.

6.022 Wernicke's area is most closely associated with:
   A. Posterior cerebral artery.
   B. Seat of intelligence.
   C. Comprehension.
   D. Reading.
   E. Expression.

C is the correct answer.
The receptive speech center.

6.023 The splenium of corpus callosum is most closely associated with:
   A. Posterior cerebral artery.
   B. Seat of intelligence.
   C. Comprehension.
   D. Reading.
   E. Expression.

A is the correct answer.
The splenium also transmits right calcarine cortex visual recordings to the receptive speech center in the left dominant hemisphere - thus also important for reading.

6.024 The association cortex is most closely associated with:
   A. Posterior cerebral artery.
   B. Seat of intelligence.
   C. Comprehension.
   D. Reading.
   E. Expression.

B is the correct answer.
The association of all varieties of sensory input with previous experience; the memory of the results of previous associations and resultant behavior; all add up to thought - ideas - intelligence.

6.025 The amygdala is most closely associated with:
   A. Primary olfactory cortex.
   B. Medial olfactory stria.
   C. Uncus of temporal lobe.
   D. Olfactory association cortex.
   E. Memory.

C is the correct answer.
(Only a small portion of the amygdala is involved with odors per se.)

6.026 The hippocampus is most closely associated with:
A. Primary olfactory cortex.
B. Medial olfactory stria.
C. Uncus of temporal lobe.
D. Olfactory association cortex.
E. Memory.

E is the correct answer.
This is old cortex, not neocortex. Bilateral hippocampal destruction results in the loss of ability to imprint and remember new information - e.g., loss of recent memory.

6.027 The parahippocampal gyrus is most closely associated with:
A. Primary olfactory cortex.
B. Medial olfactory stria.
C. Anterior perforated substance.
D. Olfactory association cortex.
E. Memory.

D is the correct answer.
Area 28 Brodmann.

6.028 The lateral olfactory stria are most closely associated with:
A. Primary olfactory cortex.
B. Medial olfactory stria.
C. Uncus of temporal lobe.
D. Olfactory association cortex.
E. Memory.

A is the correct answer.
Cortex overlying this stria forms a large part of the primary olfactory cortex. (Primary olfactory cortex also includes the anterior tip of the uncus and a small part of the amygdala.)

6.029 The septal nuclei are most closely associated with:
A. Primary olfactory cortex.
B. Medial olfactory stria.
C. Uncus of temporal lobe.
D. Olfactory association cortex.
E. Memory.

B is the correct answer.
Situated in the subcallosal area, they receive non-olfactory projections from the medial stria.

6.030 Adrenocorticotropic hormone (ACTH) is most closely associated with:
A. Premotor cortex.
B. Hypothalamus.
C. Postcentral gyrus.
D. Insula.
E. Parahippocampal gyrus.

B is the correct answer.
An anterior pituitary hormone "released" by hypothalamic hormones.

6.031 Temperature regulation is most closely associated with:
A. Premotor cortex.
B. Hypothalamus.
C. Postcentral gyrus.
D. Insula.
E. Parahippocampal gyrus.

B is the correct answer.

6.032 Pupil dilation is most closely associated with:
A. Premotor cortex.
B. Hypothalamus.
C. Postcentral gyrus.
D. Insula.
E. Parahippocampal gyrus.

B is the correct answer.
Another hypothalamic function - headquarters of the autonomic system.

6.033 Olfactory association is most closely associated with:
A. Premotor cortex.
B. Hypothalamus.
C. Postcentral gyrus.
D. Insula.
E. Parahippocampal gyrus.

E is the correct answer.

6.034 The ventral anterior nucleus thalamus is most closely associated with:
  A. Premotor cortex.
  B. Hypothalamus.
  C. Postcentral gyrus.
  D. Insula.
  E. Parahippocampal gyrus.

A is the correct answer.
VA thalamus projects to premotor cortex.

6.035 Calcarine cortex is most closely associated with:
  A. Nucleus VPL of thalamus.
  B. Medial geniculate nucleus.
  C. Nucleus ventralis lateralis of thalamus.
  D. Lateral geniculate nucleus.
  E. Anterior nucleus of thalamus.

D is the correct answer.

6.036 Somatosensory cortex is most closely associated with:
  A. Nucleus VPL of thalamus.
  B. Medial geniculate nucleus.
  C. Nucleus ventralis lateralis of thalamus.
  D. Lateral geniculate nucleus.
  E. Anterior nucleus of thalamus.

A is the correct answer.

6.037 Projection to motor cortex is most closely associated with:
  A. Nucleus VPL of thalamus.
  B. Medial geniculate nucleus.
  C. Nucleus ventralis lateralis of thalamus.
  D. Lateral geniculate nucleus.
  E. Anterior nucleus of thalamus.

C is the correct answer.

6.038 Transverse temporal gyrus of Heschl is most closely associated with:
  A. Nucleus VPL of thalamus.
  B. Medial geniculate nucleus.
  C. Nucleus ventralis lateralis of thalamus.
  D. Lateral geniculate nucleus.
  E. Anterior nucleus of thalamus.

B is the correct answer.

6.039 Receives cerebellar input. Is most closely associated with:
  A. Nucleus VPL of thalamus.
  B. Medial geniculate nucleus.
  C. Nucleus ventralis lateralis of thalamus.
  D. Lateral geniculate nucleus.
  E. Anterior nucleus of thalamus.

C is the correct answer.

6.040 Cingulate gyrus – #24 Brodmann – is most closely associated with:
  A. Nucleus VPL of thalamus.
  B. Medial geniculate nucleus.
  C. Nucleus ventralis lateralis of thalamus.
  D. Lateral geniculate nucleus.
  E. Anterior nucleus of thalamus.

E is the correct answer.

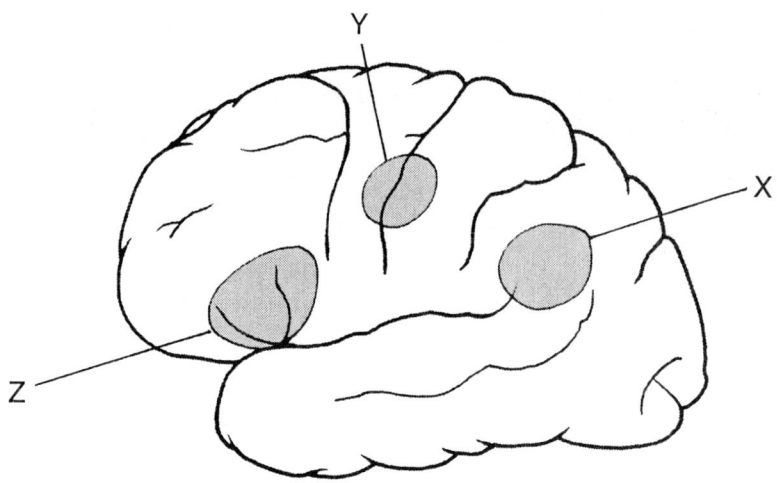

For questions 41 - 43, refer to the diagram above.

6.041 In the picture above, a lesion at "X" would produce:
  A. Agraphia, aphasia.
  B. Fluent jargon; speech comprehension poor.
  C. Fluent jargon; speech comprehension good.

B is the correct answer.
X = Receptive speech center (Wernicke)

6.042 In the picture above, a lesion at "Y" would produce:
  A. Agraphia, aphasia.
  B. Fluent jargon; speech comprehension poor.
  C. Fluent jargon; speech comprehension good.

C is the correct answer.
Y = Arcuate portion of superior longitudinal fasciculus.

6.043 In the picture above, a lesion at "Z" would produce:
  A. Agraphia, aphasia.
  B. Fluent jargon; speech comprehension poor.
  C. Fluent jargon; speech comprehension good.

A is the correct answer.
Z = Expressive speech center (Broca)

6.044 Which one of the following cortical areas is associated with olfactory sensory information?
  A. Calcarine cortex.
  B. Premotor cortex.
  C. Angular gyrus.
  D. Transverse temporal gyrus.
  E. Subcallosal cortex.

E is the correct answer.
The old subcallosal cortex with its contained septal nuclei is associated with olfactory sensory information via the medial stria of the olfactory tract. The key word here is "associated" as probably no purely odoriferous signals reach this area. These go to the primary olfactory cortex, the medial frontal cortex over and about the lateral stria and the anterior or rostral extremity of the temporal lobe's parahippocampal uncus.

6.045 The cerebral cortex can affect cerebellar activity:
  A. through the brachium pontis via corticopontine fiber tracts.
  B. via projections through the brachium conjunctivum.
  C. through a cortico-thalamo-cerebellar circuit.
  D. through the reticular-activating-system.
  E. via the ansa lenticularis.

A is the correct answer.
This is the major connection between cortex and cerebellum. Look at the size of the cerebral peduncles. These are all corticopontine save for a small central peduncular fraction of corticospinal and corticobulbar fibers. Consider the huge brachium pontis which consists entirely of pontocerebellar fibers spawned by the massive input of corticopontines to the pontine nuclei. None of the other statements make any sense!

6.046 Frontopontine fibers:
  A. enable the cerebral cortex to influence cerebellar activity.
  B. originate primarily from the pontine nuclei.
  C. course through the external capsule.
  D. project to the caudate nucleus.
  E. project bilaterally to the facial nucleus controlling the lower face.

A is the correct answer.
See previous question. Frontopontines are obviously part of the corticopontine outflow. Corticopontines come from all parts of the cerebral cortex.

6.047 Broca's area of the cerebral cortex is:
A. typically located in the right hemisphere of right-handed people.
B. the receptive area of speech.
C. located near the facial motor speech area.
D. is often damaged in cases of anterior cerebral artery occlusion.
E. directly associated with the development of visual imagery.

C is the correct answer.
Broca's area is located in the caudal inferior frontal gyrus (opercular and triangular portions) adjacent to the facial motor speech area of the precentral gyrus. Broca's is the expressive speech center. Its blood supply is from branches of the middle cerebral artery. It is practically always in the left hemisphere, though a tiny fraction of left handers house their expressive center in the right side. Fewer still use both sides in the formation of words/speech.

6.048 Lesions involving the internal capsule:
A. typically have more restricted clinical signs than focal cortical lesions.
B. can lead to upper motor neuron deficits.
C. generally cause hyporeflexia.
D. only cause motor symptoms.
E. often cause auditory loss.

B is the correct answer.
Corticospinal tracts in the internal capsule play upon and help control the activity of anterior horn cells of the spinal cord. Internal capsule lesions may interrupt these "upper motor" neuronal directions. Hyperreflexia is common. With intense compacting and consolidation of cortical inflow-outflow in the internal capsule, small lesions may produce immense sensory and/or motor deficits - far greater than those produced by small focal cortical lesions. Auditory cortical input is bilateral. Internal capsule injury will produce little or no auditory loss.

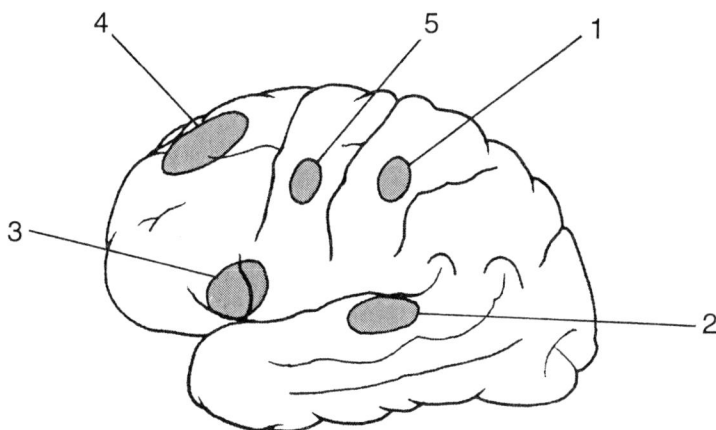

For questions 49 - 51, refer to the diagram above.

6.049 In the diagram above, which statement is INCORRECT concerning area "5" ?
A. Projects to spinal cord alpha motor neurons.
B. Contains giant pyramidal cells (Betz cells).
C. Receives input from the ventrolateral thalamic nucleus.
D. Projects corticopontine fibers to the cerebellum.
E. Is supplied by the middle cerebral artery.

D is the correct answer.
Area "5" is the forelimb area of the primary motor strip (the precentral gyrus, area 4 Brodmann). It projects to anterior horn alpha motor cells controlling forelimb muscles. This area does not send any significant fibers to the pontine nuclei. All the other statements are correct.

6.050 In the diagram above, destruction of area "3" :
A. causes Wernicke's aphasia.
B. leads to visual inability to recognize objects.
C. leads to the inability to construct the verbiage of speech.
D. destroys the emotionality of behavior.
E. destroys the supplementary motor area.

C is the correct answer.
Area "3" is the expressive speech center (Broca). Its job is to formulate words/speech in a sensible order. Wernicke's aphasia involves lesions in the temporoparietal receptive center. Area "3" is not involved with vision or emotionality. The supplementary motor area is in the medial aspect of the frontal lobe.

6.051 In the diagram above, area "2":
- A. is auditory association cortex.
- B. receives afferents from the medial geniculate body.
- C. is part of the insula.
- D. receives afferents from the lateral geniculate.
- E. is paleocortex.

A is the correct answer.
Area "2" is the auditory association area in the superior temporal gyrus, adjacent to the transverse temporal gyri of Heschl. This is perfectly good neocortex, is not part of the deeper buried insula, and has no direct projection from either geniculate nucleus.

6.052 An 80 year-old female is brought to the emergency room because she was found to be "unresponsive" in her home by neighbors. When examined, she is noted to be alert, but unable to speak other than a few short jargon words which she repeats over and over, no matter what question is asked. She cannot carry out commands. She has no weakness, but is noted on cranial nerve exam to have a right homonymous hemianopsia. You correctly locate her lesion:
- A. To the frontal lobe including Broca's area.
- B. To Wernicke's region including underlying visual radiation.
- C. To the arcuate fasciculus (superior longitudinal fasciculus).
- D. To the left temporal lobe, using the visual field defect as your indicator.
- E. To the total middle cerebral artery cortical distribution of the left.

B is the correct answer.
Wernicke's receptive speech area is involved. The patient can speak only jargon, but she can make words. Broca's motor speech area is functioning. A lesion of the arcuate fasciculus might produce jargon aphasia but the patient, with an intact receptive speech area (Wernicke) would be able to follow commands. A lesion of the left temporal lobe might produce a right visual field defect (probably less than a hemianopsia) but would not cause loss of receptive speech center activity - could follow commands. Middle cerebral artery territory includes Broca's motor speech area and the primary motor and sensory gyral strips. Paralysis and inability to make words would be found.

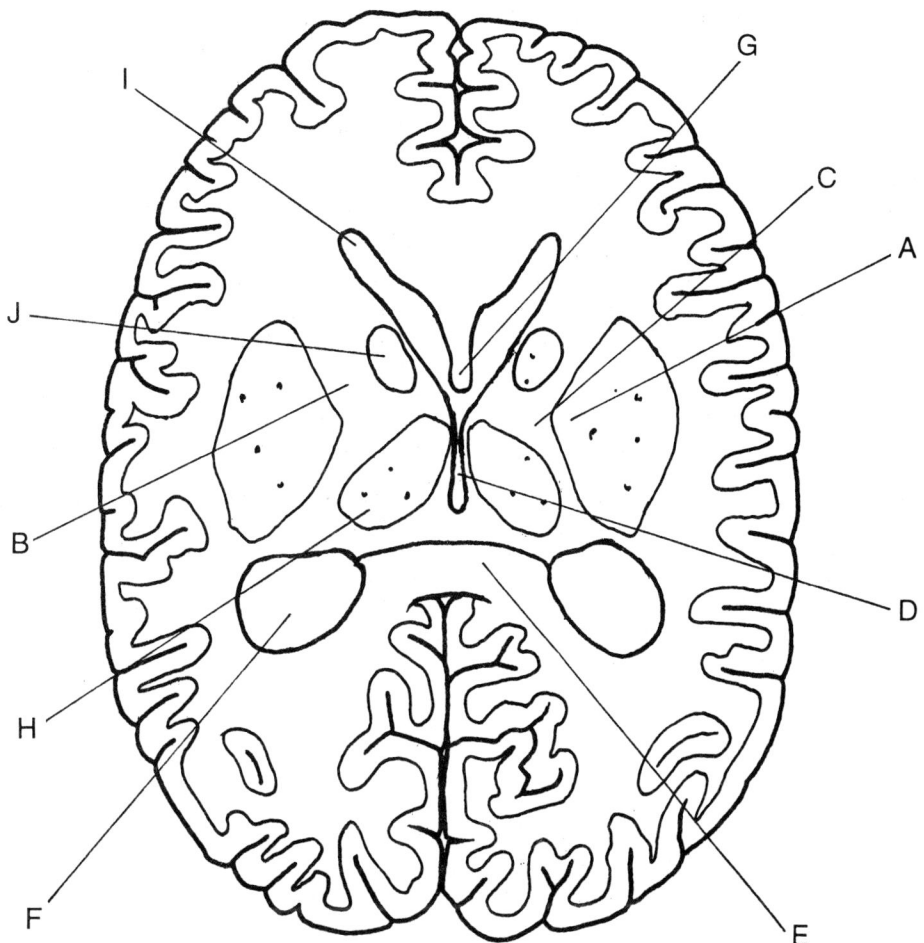

For questions 53 - 58, refer to the diagram above.

6.053 In the diagram above, _____ is the third ventricle   D is the correct answer.

6.054 In the diagram above, _____ is the body fornix   G is the correct answer.

6.055 In the diagram above, _____ are the corticobulbar fibers   C is the correct answer.

6.056 In the diagram above, _____ is the thalamus   H is the correct answer.

6.057 In the diagram above, _____ is the neostriatum   J is the correct answer.

6.058 In the diagram above, _____ is the diencephalon   H is the correct answer.

SECTION 7: CRANIAL NERVES

7.001 Which of the following statements regarding the cochlea is INCORRECT?
　A. Fluid waves in the perilymph are damped at the oval window.
　B. The cochlea consists of three channels that coil about 2-3/4 times around the modiolus.
　C. The scala media is filled with endolymph.
　D. The scalae vestibuli and tympani are filled with perilymph.
　E. The perilymph is similar to cerebrospinal fluid; is high in sodium and low in potassium.

A is the correct answer.
The statements accurately describe the cochlea except A. The fluid waves are damped at the round window. The oval footplate of the stapes fits into the oval window.

7.002 Identify one CORRECT statement concerning the auditory system.
　A. Is predominately a bilateral system.
　B. Is predominately a unilateral but crossed system.
　C. If destroyed on one side at the level of Heschl's gyrus, is associated with deafness.
　D. Has a pathway from the brainstem to the cortex which never contains more than two synapses.
　E. Utilizes the medial lemniscus as its primary ascending pathway in the brainstem.

A is the correct answer.
The auditory path has many right and left communications (at the superior olive level, between lateral lemnisci, through a commissure of the inferior colliculus). As long as stimuli are normally perceived and transported to the cochlear nuclei, the system is bilateral. Hence unilateral pathway lesions will not result in deafness. Both sides of the auditory cortex hear the same thing.

7.003 Which one of the following is not involved in the auditory pathway?
　A. Nucleus of the lateral lemniscus.
　B. Brachium of the inferior colliculus.
　C. Inferior olive.
　D. Medial geniculate.
　E. Trapezoid body.

C is the correct answer.
The superior olive is prominently involved in audition. The inferior olive could not care less!

7.004 Which of the following structures is not associated with the auditory pathway?
　A. Inferior colliculus.
　B. Sublenticular portion of the internal capsule.
　C. Lateral lemniscus.
　D. Lateral geniculate nucleus of the thalamus.
　E. Transverse temporal gyrus.

D is the correct answer.
The lateral geniculate nucleus of the thalamus is concerned with optical matters. All the other statements are accurate.

7.005 Select the one CORRECT statement regarding the auditory pathway.
　A. Projects to the primary auditory cortex (Heschl's gyrus) bilaterally.
　B. Is a predominately ipsilateral system.
　C. Follows the lateral lemniscus to the lateral geniculate.
　D. If interrupted by a lesion of the superior olive will produce ipsilateral deafness.
　E. Involves the inferior colliculus and lateral geniculate body on its way to the cortex.

A is the correct answer.
After the above questions concerning the auditory path, this one should be easy.

7.006 Select the one CORRECT statement regarding the oculomotor nerve.
 A. Arises in the upper pons.
 B. Innervates branchiogenic musculature of the orbit.
 C. Decussates in the midbrain area.
 D. Travels with the optic nerve through the optic canal to reach the orbit.
 E. Receives input originating in the cerebral cortices.

E is the correct answer.
Oculomotor nerve arises in the midbrain and innervates the somitic extraocular muscles. It does not decussate. It enters the orbit through the superior orbital fissure. Cerebral cortical directions via corticobulbar fibers obviously control volitional eye movements and the fast checking movement of nystagmus - all involving the oculomotor nerve.

7.007 Which one of the following statements regarding the spinal tract of the trigeminal nerve is INCORRECT?
 A. Intermingles with the auditory fibers in the lower pons to form the trapezoid body.
 B. Contains fibers carrying pain sensations from the larynx.
 C. Extends from midbrain to cervical spinal levels.
 D. Synapses in the neighboring nucleus of the spinal tract, proceeds to the contralateral ventral posteromedial (VPM) nucleus of the thalamus.
 E. Is entirely homologous to Lissauer's tract in the spinal cord.

A is the correct answer.
The medial lemniscus is the tract intermingling with the trapezoid body in the pons.

7.008 Which one of the following is an SVE nucleus?
 A. Dorsal motor nucleus of the vagus.
 B. Nucleus ambiguus.
 C. Abducens nucleus.
 D. Hypoglossal nucleus.
 E. Trochlear nucleus.

B is the correct answer.
SVE nuclei innervate musculature arising in the branchial arches. Extraocular muscles and the muscles of the tongue arise from somites. Dorsal motor nucleus of vagus innervates smooth muscle, glands, and cardiac muscle.

7.009 One of the following statements regarding the dorsal motor nucleus of the vagus is INCORRECT.
 A. Receives afferents from the anterior hypothalamic nuclei.
 B. Is located in the floor of the fourth ventricle.
 C. Controls the muscles of the jejunum.
 D. Controls the muscles of the pharynx.
 E. Contains neurons whose axons synapse in neurons of the myenteric plexus.

D is the correct answer.
Dorsal motor nucleus is parasympathetic headquarters of the vagus, receiving direction originating in parasympathetic nuclei located in the anterior hypothalamus. It rests in the floor of the fourth ventricle. It controls smooth muscle of the digestive tract as far down as the splenic flexure area, after synapse in postganglionic neurons of the myenteric and submucosal plexi. Pharyngeal muscle is controlled via the nucleus ambiguus - the major branchial muscle nucleus.

7.010 Select the one best answer. The trochlear nerve:
 A. crosses to the opposite side after leaving the nucleus.
 B. leaves from the nucleus nestled under the superior colliculus.
 C. Contains autonomic fibers innervating the sphincter of the iris.
 D. is uncrossed.
 E. innervates the inferior oblique muscle.

A is the correct answer.
Trochlear nerve nucleus is deep to the inferior colliculus. It crosses in the superior medullary velum and innervates the superior oblique muscle. It carries no autonomic fibers.

7.011 The tenth cranial nerve (vagus) would have which one of the following nuclei associated with it?
 A. Inferior salivatory nucleus.
 B. Nucleus solitarius.
 C. Superior salivatory.
 D. Inferior olivary nucleus.
 E. Accessory cuneate nucleus.

B is the correct answer.
Afferents from the viscera innervated by the vagus (other than pain) are conveyed to the nucleus solitarius. (Pain is apparently carried in sympathetic fibers which retrace sympathetic routes to enter the cord and proceed to consciousness via the spinothalamic path.)

7.012 Which one of the following statements correctly describes the innervations of the oculomotor nerve?
   A. The medial, superior, and inferior rectus muscles, the inferior oblique, levator palpebrae, and the iris constrictor.
   B. The medial, superior, and inferior rectus muscles, the superior oblique, levator palpebrae, and the iris constrictor.
   C. The lateral, superior, and inferior rectus muscles, the superior oblique, levator palpebrae, and the iris constrictor.
   D. The orbicularis oculi, in addition to other extraocular muscles and the iris constrictor.
   E. All muscles producing extraocular movement.

A is the correct answer.
The oculomotor nerve does not innervate the superior oblique or the lateral rectus.

7.013 Concerning the trigeminal spinal nucleus, all the following statements are true EXCEPT:
   A. Occupies the same relative position in the brainstem tegmentum as does the substantia gelatinosa in the cord.
   B. Extends from the pons to the upper cervical segments of the cord.
   C. Projects to the ipsilateral VPM.
   D. Is connected with neurons in the semilunar ganglion.
   E. Serves modalities of pain and temperature.

C is the correct answer.
Projection of the spinal nucleus of the trigeminal is the same as the projection of neurons of the substantia gelatinosa of the cord, i.e., to the contralateral VPM.

7.014 Which statement regarding the trochlear nerve is CORRECT?
   A. It conveys postganglionic sympathetic fibers.
   B. It stems from a nucleus deep to the inferior colliculus.
   C. It emerges from the mesencephalon through the interpeduncular fossa.
   D. It decussates in the posterior commissure.
   E. It enters the orbit through the inferior orbital fissure.

B is the correct answer.
The trochlear nerve carries no autonomic fibers. (These are all in the oculomotor nerve.) The trochlear is the only cranial nerve emerging from the dorsal brainstem. It decussates in the superior medullary velum, before encircling the cerebral peduncles to enter the cavernous sinus with the other extraocular eye muscle nerves. They enter the orbit through the superior orbital fissure.

7.015 Which statement regarding the gag reflex is CORRECT?
   A. It is strictly unilateral (i.e., direct, without a consensual response).
   B. Is composed of a trigeminal afferent limb and a vagus-glossopharyngeal efferent limb.
   C. Is typically lost with unilateral motor cortex lesions.
   D. Is composed of a vagus-glossopharyngeal (X-IX) afferent limb, and a vagus-glossopharyngeal (X-IX) efferent limb.
   E. Involves both the pons and the medulla.

D is the correct answer.
The gag reflex is a consensual response relying on the integrity of the IX and X nerves. Classically, it is taught that IX is the afferent limb, and X the efferent limb. These two nerves, however, are so intermingled in the innervation of this area, it is impossible to make responsibility so discrete. Loss of cortex and/or pons should not affect the reflex as IX and X are medullary-based and bilaterally innervated.

7.016 Ice water irrigation of the right ear:
  A. Will produce nystagmus with the slow component to the right.
  B. May result in frostbite of the tympanum.
  C. May produce diplopia.
  D. Depends on a vagus afferent limb and a glossopharyngeal afferent limb to produce noticeable effects.
  E. Will produce eye motions as long as the brainstem is intact.

A is the correct answer.
At rest, the right and left horizontal canals demonstrate a basic tonic firing and "cancel each other out". Cooling the right horizontal canal causes a convection current flow downward and away from the canal ampulla. This results in a decrease in the firing rate of the irrigated canal. The now unopposed unirrigated left canal's tonic firing rate will signal an apparent horizontal rotatory movement to its side; the eyes will be driven to the right. This is the slow movement of nystagmus. The opposite "correcting" rapid movement depends on an intact cerebral cortex and yoking PPRF-MLF system. With decortication, deviation alone will be seen; in this case to the right. Deviation will be seen as long as the systems "below" the cortex are intact. Diplopia is not present with normal eye muscle innervation. Frostbite!! Report it!!

7.017 Select one CORRECT statement. Nystagmus with the slow component to the right will occur with:
  A. Cold water irrigation of the left ear.
  B. Destruction of the left VIII nerve by an acoustic neuroma.
  C. Sudden stop after a prolonged spinning movement to the right.
  D. Destruction of the PPRF on the left.
  E. Oculomotor nerve dysfunction.

C is the correct answer.
Nystagmus with slow component to the right occurs with stimulation (hot water irrigation) of the left vestibular apparatus. Cessation of a right spin will stimulate the left side due to continued inertial flow pressure on the left. (Inertial flow will be away from the right ampulla.) Cold water irrigation of the left ear results in relative overactivity on the right, with consequent nystagmus with slow component to the left. Similarly, destruction of the left VIII nerve will result in apparent stimulation of the right side and nystagmus with slow component to the left. Destruction of the left PPRF will eliminate any lateral gaze movement to the left. Oculomotor nerve dysfunction will prevent nystagmal adduction of the eye on the involved side.

7.018 Identify the one best answer below regarding the vestibular system.
  A. Has little contribution to extraocular muscle reflex activity.
  B. Distributes to the conjugate eye systems to control reflex adjustments of the eyes during head movement.
  C. Innervates the fast movement of nystagmus, the slow movement being a function of the cerebral hemispheres.
  D. When disordered, is rarely associated with nystagmus.
  E. Is a vestigial functional system in primates, the eyes having taken over virtually all equilibrium functions.

B is the correct answer.
This is one of the cardinal functions of the vestibular system. In adjusting the eyes to head movement, a slow coincidental eye movement will occur, orchestrated by the vestibular-eye yoking systems, to maintain visual fixation. With nystagmus, the slow eye movement of vestibular origin persists. The fast "correcting" movement is a function of the cerebral hemispheres. Nystagmus is a frequent accompaniment of vestibular disorder due to "imbalance" of the right and left canals. The eyes help with balance, but do not have the rapid reflex adjustments accompanying vestibular stimulation. Also, the eyes are not much good in the dark.

7.019 Concerning vertigo, which statement is CORRECT?
  A. A synonym for nausea.
  B. The sensation of movement of self (subject) or the environment (object).
  C. Caused by disorder of cerebellar afferent tracts.
  D. Perceived as a moving to and fro (two directions) of the environment or self.
  E. A problem in the weightless state.

B is the correct answer.
Vertigo may be associated with nausea, but no necessarily so. The cerebellum is not vertignous.

7.020 Only one statement below regarding the sense of smell is CORRECT.
  A. Can be decreased by a unilateral lesion anywhere from the olfactory bulb to the olfactory cortex of the temporal lobe.
  B. Is dependent upon the presence of a relatively dry mucous membrane in the roof of the nasal passage.
  C. Is an excellent system for the localization of odors.
  D. Is most often disordered in the event of extradural hematoma.
  E. Is important for the recognition of some potentially dangerous volatile substances.

E is the correct answer.
Despite its reduced size in the human, olfactory sense is remarkable in its ability to recognize minute amounts of volatile substance. It is not good in determining the origin of odors. To produce frontal hyposmia, a unilateral lesion must involve structures rostral to the olfactory trigone, i.e., olfactory tract, olfactory bulb, nerve filaments, or the olfactory nasal mucosa. Otherwise, generous interconnections via the anterior commissure deliver unilateral olfactory stimuli to both temporal olfactory cortices. Volatile molecules must be dissolved in nasal mucous in order to be recognized by nasal olfactory receptors. Olfaction is often disordered by occipital head trauma; the brain moving backward and shearing off the olfactory nerve fibers as they pass through the cribriform plate. Epidural hematoma will not have this effect.

7.021 Involvement of one nucleus ambiguus by a destructive lesion (e.g., a stroke) will be associated with one of the following:
  A. Difficulty swallowing, hoarseness, and weakness of elevation of the ipsilateral shoulder and of turning the head contralaterally.
  B. Hypotension and bradycardia.
  C. Hypomotility of the bowel.
  D. Difficulty swallowing and decreased taste over the posterior third of the tongue.
  E. Difficulty swallowing and hoarseness.

E is the correct answer.
Ambiguus is involved with function of the branchogenic muscles. Injury, therefore, will result in difficulty in swallowing and hoarseness. The sternocleidomastoid and trapezius muscles are branchogenic but are driven by the spinal nucleus of the spinal accessory nerve. Hypotension and bradycardia are in autonomic territory, as is motility of the bowel. Ambiguus is not involved in sensory input.

7.022 Which of the following components is not present in the facial nerve?
  A. SSA.
  B. GVE.
  C. SVA.
  D. SVE.
  E. GSA.

A is the correct answer.
A Special Somatic Afferent component describes outside influences affecting the eyes and ears - sight, sound, balance. The facial nerve carries General Visceral Efferent fibers - greater petrosal and chorda tympani; Special Visceral Afferents - taste in chorda tympani; Special Visceral Efferent fibers to the branchogenic muscles of facial expression, and a tiny area of ear skin carried by an auricular branch which is mainly concerned with innervating the ear wiggling muscles.

7.023 A patient with Bell's peripheral palsy of the facial nerve could have any one of the following EXCEPT:
  A. hyperacusis.
  B. impaired taste sensation.
  C. impaired sense of smell.
  D. paralysis of the upper face.
  E. paralysis of the lower face.

C is the correct answer.
The facial nerve is ubiquitous but never ventures into the olfactory department. Hyperacusis may occur due to paralysis of the facial nerve innervated tensor tympani muscle. Taste in the anterior two-thirds of the tongue is carried in the chorda tympani.

7.024 Taste from the anterior two thirds of the tongue:
  A. is associated with nucleus solitarius only.
  B. is associated with nucleus of spinal tract V only.
  C. is associated with both nucleus solitarius and nucleus of spinal tract V.
  D. is associated with neither nucleus solitarius nor nucleus of spinal tract V.

A is the correct answer.
Cranial end of the solitarius may be called the "gustatory center". All taste fibers register here.

7.025 Pain of a toothache:
  A. is associated with nucleus solitarius only.
  B. is associated with nucleus of spinal tract V only.
  C. is associated with both nucleus solitarius and nucleus of spinal tract V.
  D. is associated with neither nucleus solitarius nor nucleus of spinal tract V.

B is the correct answer.
Alveolar nerves of V.

7.026 Salivary secretion:
  A. is associated with nucleus solitarius only.
  B. is associated with nucleus of spinal tract V only.
  C. is associated with both nucleus solitarius and nucleus of spinal tract V.
  D. is associated with neither nucleus solitarius nor nucleus of spinal tract V.

D is the correct answer.
Salivary secretion is ordered by the salivatory nucleus.

7.027 Carotid body chemoreception:
  A. is associated with nucleus solitarius only.
  B. is associated with nucleus of spinal tract V only.
  C. is associated with both nucleus solitarius and nucleus of spinal tract V.
  D. is associated with neither nucleus solitarius nor nucleus of spinal tract V.

A is the correct answer.
GVA of IX.

7.028 Laryngeal pain:
  A. is associated with nucleus solitarius only.
  B. is associated with nucleus of spinal tract V only.
  C. is associated with both nucleus solitarius and nucleus of spinal tract V.
  D. is associated with neither nucleus solitarius nor nucleus of spinal tract V.

B is the correct answer.
Coming from the larynx, these pain fibers may be listed as "GVA" though in reality they should be GSA as they carry pain-temperature-touch to the trigeminal system.

7.029 Satiety:
  A. is associated with nucleus solitarius only.
  B. is associated with nucleus of spinal tract V only.
  C. is associated with both nucleus solitarius and nucleus of spinal tract V.
  D. is associated with neither nucleus solitarius nor nucleus of spinal tract V.

A is the correct answer.
This type of sensation is truly visceral.

7.030 Pain from the posterior third of the tongue:
  A. is associated with nucleus solitarius only.
  B. is associated with nucleus of spinal tract V only.
  C. is associated with both nucleus solitarius and nucleus of spinal tract V.
  D. is associated with neither nucleus solitarius nor nucleus of spinal tract V.

B is the correct answer.
"Ordinary" sensation here is carried by IX to the trigeminal apparatus.

7.031 Light reflex is most closely associated with:
  A. II and III cranial nerves.
  B. V.
  C. X.
  D. IX and X.
  E. V and VII.

A is the correct answer.
Requires an intact retina/optic nerve, and parasympathetics of III.

7.032 Gag reflex is most closely associated with:
  A. II and III cranial nerves.
  B. V.
  C. X.
  D. IX and X.
  E. V and VII.

D is the correct answer.
Sensory arm is IX, motor arm is X.

Cranial Nerves

7.033 Jaw jerk is most closely associated with:
  A. II and III cranial nerves.
  B. V.
  C. X.
  D. IX and X.
  E. V and VII.

B is the correct answer.
Tapping the point of the jaw produces a stretch reflex shortening of the jaw closers. Sensory and motor arms are V.

7.034 Accommodation reflex is most closely associated with:
  A. II and III cranial nerves.
  B. V.
  C. X.
  D. IX and X.
  E. V and VII.

A is the correct answer.

7.035 Corneal reflex is most closely associated with:
  A. II and III cranial nerves.
  B. V.
  C. X.
  D. IX and X.
  E. V and VII.

E is the correct answer.
Sensory arm is from ophthalmic division of V - ciliary nerves; motor arm is VII.

7.036 Which of the following occurs in all peripheral and facial palsies?
  A. Hyperacusis.
  B. Corneal reflex.
  C. Corneal drying and ulceration.
  D. Flattening of nasolabial fold.
  E. Diminished auditory acuity.

D is the correct answer.

7.037 Paralysis of stapedius muscle is most closely associated with:
  A. Hyperacusis.
  B. Corneal reflex.
  C. Corneal drying and ulceration.
  D. Flattening of nasolabial fold.
  E. Diminished auditory acuity.

A is the correct answer.

7.038 Which of the following is a complication of peripheral facial paralysis?
  A. Hyperacusis.
  B. Corneal reflex.
  C. Corneal drying and ulceration.
  D. Flattening of nasolabial fold.
  E. Diminished auditory acuity.

C is the correct answer.

7.039 Which of the following is present despite central type of facial paralysis?
  A. Hyperacusis.
  B. Corneal reflex.
  C. Corneal drying and ulceration.
  D. Flattening of nasolabial fold.
  E. Diminished auditory acuity.

B is the correct answer.
A central type of facial palsy, due to bilaterality of innervation of the upper face, leaves the frontalis and orbicularis oculi muscles intact while the lower face is paralyzed. Corneal reflex will persist, the eye can be closed. Peripheral facial palsy finds the entire ipsilateral side of the face immobile, the eye cannot be closed and corneal drying and ulceration threaten. In both types of facial palsy, the nasolabial fold will be flattened due to loss of facial muscle tone. Hyperacusis is the result of paralysis of the sound-damping stapedius muscle which is innervated by VII. Hearing is not otherwise affected.

7.040 Taste is most closely associated with:
A. Geniculate ganglion.
B. Sphenopalatine ganglion.
C. Ciliary ganglion.
D. Otic ganglion.
E. Sympathetic trunk.

A is the correct answer.
Chorda tympani – VII. The others are autonomic ganglia.

7.041 Cranial nerve IX is most closely associated with:
A. Geniculate ganglion.
B. Sphenopalatine ganglion.
C. Ciliary ganglion.
D. Otic ganglion.
E. Sympathetic trunk.

D is the correct answer.
Parotid innervation; lesser petrosal of IX.

7.042 The Edinger Westphal nucleus is most closely associated with:
A. Geniculate ganglion.
B. Sphenopalatine ganglion.
C. Ciliary ganglion.
D. Otic ganglion.
E. Sympathetic trunk.

C is the correct answer.
Preganglionic parasympathetics to the ciliary ganglion.

7.043 Pupillary constriction is most closely associated with:
A. Geniculate ganglion.
B. Sphenopalatine ganglion.
C. Ciliary ganglion.
D. Otic ganglion.
E. Sympathetic trunk.

C is the correct answer.
A parasympathetic function. (Sympathetics innervate the iris dilator.)

52 Cranial Nerves

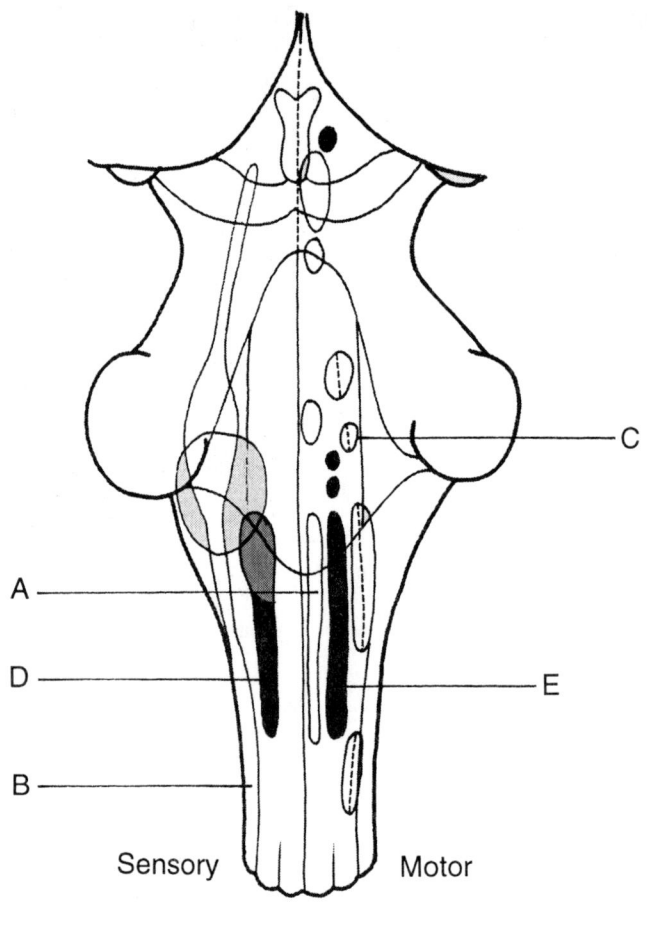

Cranial nerve nuclei

For questions 44 - 49, refer to the diagram above.

7.044 In the diagram above, which cranial nerve nucleus is most closely associated with the urge to urinate?

D is the correct answer.
Nucleus solitarius.

7.045 In the diagram above, which cranial nerve nucleus is most closely associated with an aching tooth?

B is the correct answer.
Nucleus - spinal tract V.

7.046 In the diagram above, which cranial nerve nucleus is most closely associated with intestinal peristalsis?

E is the correct answer.
Dorsal motor nucleus vagus.

7.047 In the diagram above, which cranial nerve nucleus is most closely associated with the orbicularis oculi?

C is the correct answer.
Facial nucleus.

7.048 In the diagram above, which cranial nerve nucleus is most closely associated with gastric acid secretion?

E is the correct answer.
Dorsal motor nucleus vagus.

7.049 In the diagram above, which cranial nerve nucleus is most closely associated with whistling?

C is the correct answer.
Facial nucleus.

7.050 Your patient's face is weak on the right. She complains of pain behind the ear and difficulty with taste.
- A. You would not be surprised if she also complains that loud low frequency noises are bothersome on the right.
- B. It stands to reason that only the lower part of her face will be involved.
- C. You should expect some long tract signs - e.g. a hemiparesis or ataxia.
- D. You would assume the brainstem is involved because the various functions of the facial nerves are found in a single locus.
- E. None of the above.

A is the correct answer.

The facial nerve is obviously involved. Tiny branches of the facial nerve innervate the skin over the mastoid process (and the external auditory meatus). The facial nerve also innervates the stapedius muscle which helps control ossicle induced wave amplitude in the inner ear. The various functions of the facial nerve are not in a single locus in the brainstem. Sensory fibers in the mastoid skin travel to the spinal tract of the trigeminal; those from the taste buds of the tongue to the nucleus solitarius; motor fibers from the facial nucleus proper. Central lesions involving corticobulbar fibers to the facial nucleus result in lower facial palsy. However, such a lesion would not produce the sensory symptoms found here. The best explanation for the symptoms described here is a peripheral facial nerve lesion affecting the peripheral nerve central to the geniculate ganglion. Long tracts need not be involved. The facial paralysis will involve the entire face.

SECTION 8: CSF & MENINGES

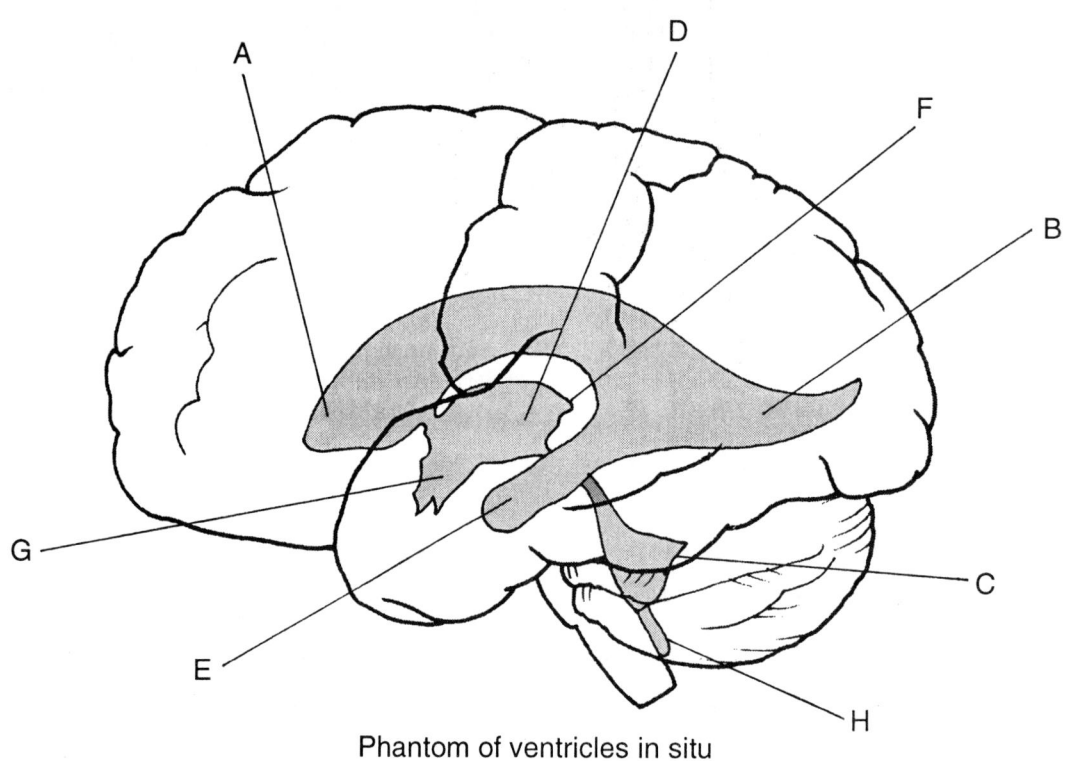

Phantom of ventricles in situ

For questions 1 - 4, refer to the diagram above.

8.001 In the diagram above, ____ contains the choroid plexus of the lateral ventricle.

E is the correct answer.

8.002 In the diagram above, ____ is an area of communication of the ventricular system with the subarachnoid space.

C is the correct answer.

8.003 In the diagram above, ____ is the location of the choroid plexus of the fourth ventricle.

H is the correct answer.

8.004 In the diagram above, ____ is the location of the pineal recess.

F is the correct answer.

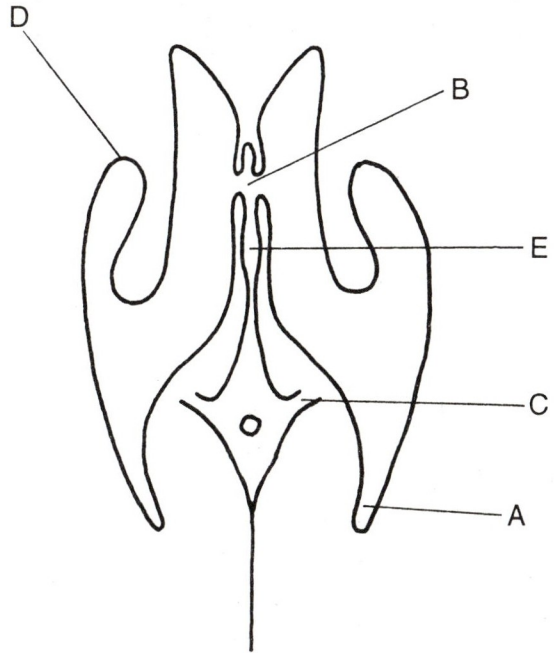

For questions 5 - 8, refer to the diagram above.

8.005 In the diagram above, ____ is the third ventricle.   E is the correct answer.

8.006 In the diagram above, ____ is the inferior horn lateral ventricle.   D is the correct answer.

8.007 In the diagram above, ____ is the interventricular foramen.   B is the correct answer.

8.008 In the diagram above, ____ is the foramen of Luschka.   C is the correct answer.

8.009 Increased intracranial pressure will result from blockage of the foramina of the: D is the correct answer.
   A. Epidural space.
   B. Subarachnoid space.
   C. Subdural space.
   D. Fourth ventricle.
   E. Lumbar cistern.

8.010 At spinal levels, which of the following contains the internal vertebral venous plexus?   A is the correct answer.
   A. Epidural space.
   B. Subarachnoid space.
   C. Subdural space.
   D. Fourth ventricle.
   E. Lumbar cistern.

8.011 Which of the following is a potential space where venous blood may accumulate to produce a hematoma?   C is the correct answer.
Epidural hematomas are classically arterial in origin.
   A. Epidural space.
   B. Subarachnoid space.
   C. Subdural space.
   D. Fourth ventricle.
   E. Lumbar cistern.

CSF & Meninges

8.012 Bleeding from branches of the middle meningeal artery may create an actual space from potential space:
A. Epidural space.
B. Subarachnoid space.
C. Subdural space.
D. Fourth ventricle.
E. Lumbar cistern.

A is the correct answer.

8.013 The lamina terminalis is most closely associated with:
A. Lateral ventricle.
B. Third ventricle.
C. Fourth ventricle.
D. Cerebral aqueduct.
E. Foramen of Luschka.

B is the correct answer.

8.014 The hypothalamus is most closely associated with:
A. Lateral ventricle.
B. Third ventricle.
C. Fourth ventricle.
D. Cerebral aqueduct.
E. Foramen of Luschka.

B is the correct answer.

8.015 The subarachnoid space is most closely associated with:
A. Lateral ventricle.
B. Third ventricle.
C. Fourth ventricle.
D. Cerebral aqueduct.
E. Foramen of Luschka.

E is the correct answer.

8.016 The caudate nucleus is most closely associated with:
A. Lateral ventricle.
B. Third ventricle.
C. Fourth ventricle.
D. Cerebral aqueduct.
E. Foramen of Luschka.

A is the correct answer.

8.017 Concerning the dura mater, one of the following is INCORRECT:
A. It is the only one of the meninges capable of responding to a painful stimulus.
B. It is nourished by blood vessels that also nourish the cranial bones.
C. Demonstrates a horizontally disposed fold that separates the middle from the posterior cranial fossa.
D. Is pierced by the internal carotid artery at the foramen lacerum.
E. Is tightly applied to the overlying cranial bones.

D is the correct answer.
The internal carotid pierces the dura just posterior to the anterior clinoid, after traversing the cavernous sinus. The foramen lacerum is a "pseudo" foramen seen only in the dry skull, at the apex of the petrous pyramid of the temporal bone. It appears as a jagged opening in the bone about the carotid canal. In life, this opening is filled with cartilage, completing the covering of the artery in its canal. No structure passes through this "foramen". All the other statements are true. The horizontally disposed fold is the tentorium cerebelli.

8.018 Find the INCORRECT statement regarding the pia mater:
A. It is a delicate vascular layer.
B. It gives rise to tethering ligaments of the spinal cord.
C. It extends into sulci and fissures.
D. It forms part of the tela choroidea of the midbrain.
E. It does not permit passive transudation from capillaries to the cerebral extracellular fluid.

D is the correct answer.
The pia does form part of the tela choroidea of ventricles into which choroidal plexus capillaries invaginate. But there is no ventricle in the midbrain, no choroid plexus, no tela choroidea. Tethering pial ligaments are the denticulate and pial filum terminale. All statements but D accurately describe the pia.

8.019 One CORRECT statement regarding the subarachnoid space is contained in the suggestions below:
A. Communicates with the spinal cord subarachnoid space through the straight sinus.
B. May be pierced and entered safely by a needle inserted at the level of the cauda equina.
C. Is only a potential space. It appears, filled with blood, only after skull fracture.
D. Forms arachnoid villi which are necessary for the absorption of nutrients from the blood.
E. Extends to the level L1-2.

B is the correct answer.
The subarachnoid space is very real space, filled with cerebrospinal fluid. The familiar spinal tape is performed at cauda equina levels to avoid injury to the spinal cord proper. The spinal subarachnoid space is a continuation of the subarachnoid space about the brain. The straight venous sinus has no connection, With any trauma, stroke, etc., the injured brain vessels may bleed into the subarachnoid space. Normally, there are no blood cells in the CSF. The arachnoid villi are necessary only for the transfer of CSF to the venous blood. The subarachnoid space extends to the level of S2 where the dural/arachnoidal tube ends.

8.020 Regarding the cranial epidural space, one of the following is INCORRECT:
A. May be the site of post-traumatic arterial hemorrhage.
B. Unlike its counterpart in the spine is a potential space.
C. Contains arachnoid granulations.
D. Contains a branch of the maxillary artery.
E. Is traversed by emissary veins.

C is the correct answer.
Arachnoid granulations invaginate the dural sinuses, but do not extend into the epidural space. The latter is a potential space between dura and the calvaria. The latter two are hard to separate. In the spine the dura is not tightly applied to the surrounding bone, and a real space exists which houses the venous internal vertebral plexus. The middle meningeal artery, a branch of the maxillary, enters the interior of the skull via the foramen spinosum, and courses in the epidural "space" in bony grooves. Skull fracture through a meningeal bony groove will likely rupture the contained vessel. Arterial bleeding is powerful enough to force open the potential epidural space to form an expanding dangerous epidural hematoma. Emissary veins cross the dura-calvarial junction in their course between dural sinus and scalp veins.

8.021 Find the one CORRECT answer in the statements below regarding the cerebrospinal fluid:
A. Normally the cerebrospinal fluid contains 5 - 10 WBC per ml.
B. CSF is produced in large part by diffusion through the walls of the capillaries of the choroid plexi and ventricular ependyma.
C. The total volume in a 70 kg man is approximately 400 ml.
D. CSF is normally sterile.
E. CSF has approximately the same electrolyte composition as blood serum.

D is the correct answer.
Finding of bacteria in a CSF tap indicates infection somewhere in the CNS. CSF is normally sterile and free of cells. The production of CSF is an energy dependent process, not a simple diffusion. In comparison to blood, it is higher in Na, Cl, and Mg; lower in K and Ca. The total volume of CSF in an adult is about 150 ml.

8.022 One statement below accurately places the lumbar cistern:
A. Epidural space.
B. Subarachnoid space.
C. Subdural space.
D. Space between the ependyma and the pia.
E. Central canal of the spinal cord.

B is the correct answer.
Epidural and subdural spaces are potential spaces. The ependyma lines the ventricles, the pia the external surface of the CNS. Never the twain shall meet. The central canal is tiny, often occluded - vestigial. The lumbar cistern is the relatively large subarachnoid space between the conus medullaris at L1-2 and the caudal extent of the dural arachnoid sac at S2.

8.023 Select the one INCORRECT statement regarding the meninges and the cerebrospinal fluid:
- A. Inflammatory meningitis may obstruct the flow of CSF producing hydrocephalus and increased intracranial pressure.
- B. Of the three meningeal layers, only the dura, innervated by the trigeminal and upper cervical nerves, is pain sensitive.
- C. The choroid plexi secrete all of the CSF.
- D. Lumbar puncture in the presence of increased intracranial pressure may result in herniation of cerebellar tonsils through the foramen magnum with resultant pressure on the medulla and sudden death.
- E. Normally the ventricles and the subarachnoid space contain about 150 ml of CSF.

C is the correct answer.
About one third of the CSF is secreted via the cerebral capillaries. As with the choroid plexus, this is an energy dependent eclectic transfer, not a simple diffusion. All the other statements are accurate.

8.024 Concerning the cerebrospinal fluid, one of the statements below is CORRECT:
- A. Is secreted at a volume of 500 ml per 24 hours.
- B. Contains protein and glucose in approximately equal amounts per ml as in the serum.
- C. Is absorbed in part by the emissary veins.
- D. Is formed by the choroid plexus of the mesencephalic aqueduct.
- E. Fills the subarachnoid space from the region of the dome of the cerebral hemispheres to the level of the L2 vertebral segment.

A is the correct answer.
The secretion of CSF results in a turnover of about 500 ml per 24 hours. There is a huge difference in protein concentration between blood serum (6 - 8 gm/dl) and CSF (15 - 40 mgm/dl). Glucose concentration in the CSF is about half that in the serum. Emissary veins are not in contact with CSF as they run through the crania and scalp. There is no choroid plexus in the aqueduct. The extent of the CSF-filled subarachnoid space is from cerebral dome to the level of S2 vertebra.

8.025 One statement regarding the location of the choroid plexus is INCORRECT:
- A. The interventricular foramina.
- B. The inferior horn of the lateral ventricle.
- C. The third ventricle.
- D. The caudal part of the fourth ventricle.
- E. The aqueduct of Sylvius.

E is the correct answer.
Choroid plexus is present in the roof of the third ventricle, through the interventricular foramina and around the "C" of the lateral ventricles to the tips of their inferior (temporal) horns. Choroid plexus is also present in the roof of the fourth ventricle in its caudal part. There is no choroid plexus in the midbrain.

8.026 In a successful spinal tap for cerebrospinal fluid performed in the midline of the L3-L4 interspace, which one of the following structures would be punctured by the spinal tap needle?
- A. The anterior longitudinal ligament.
- B. Denticulate ligament.
- C. Arachnoid membrane.
- D. Pia mater.
- E. The central canal of the cord.

C is the correct answer.
Anterior and posterior longitudinal ligaments are applied to the anterior and posterior surfaces of the vertebral bodies. The spinal canal is posterior to the posterior longitudinal ligament. The cord pia and the denticulate ligament extend only to the end of the cord at L1-L2 level. The tap is done through L3-4 (or L4-5) interspace. Obviously, the arachnoid must be punctured to obtain fluid from the subarachnoid space. The central canal of the cord is tiny, vestigial, often dry.

8.027 Circulation of the cerebrospinal fluid is dependent on all but one of the following:
- A. Unobstructed flow of CSF from the lateral ventricles to the cisterna magna.
- B. Passage of the CSF through arachnoid granulations.
- C. Continuous section of CSF by the choroid plexi.
- D. Patent foramina of Luschka and Magendie.
- E. Patency of the central canal of the cord.

E is the correct answer.
The vestigial central canal of the cord is tiny, often occluded, and not a significant player in CSF circulation. Circulation depends on free communication between ventricular and subarachnoid spaces through patent foramina of Luschka and Magendie; and functioning arachnoid villi.

8.028 Which one of the following statements is FALSE?
A. The CSF is largely derived from the choroid plexus.
B. The content of the CSF is secreted by the choroid epithelium.
C. The CSF and the extracellular fluid of the brain substance are in free communication.
D. The CSF forms a buoyant fluid medium which floats the entire central nervous system.
E. The extracellular fluid of the brain is a transudate, similar to that formed in any other capillary bed of the body.

E is the correct answer.
As opposed to other body areas, the brain requires energy to transport substances from blood to brain extracellular fluid through the tight junctions of the cerebral capillary epithelium. About two-thirds of the CSF is secreted by the choroid epithelium underlying the choroid capillary plexus. The ECF and CSF are indeed in free communication. The CSF forms a fluid jacket about the entire CNS.

8.029 One statement regarding the lumbar cistern is INCORRECT:
A. Contains no neuron cell bodies.
B. Contains CSF which is distinctly different from the CSF surrounding the brain proper in its chemical makeup.
C. Extends from the tip of the conus medullaris to the level of the S2 vertebral segment.
D. Is traversed by nerve fibers of the cauda equina.
E. Is traversed by the pial filum terminale.

B is the correct answer.
Unless blocked by some disease process, the CSF of the lumbar cistern is of exactly the same chemical makeup as CSF elsewhere. The cauda equina and the pial filum terminale run through the cistern which extends from the end of the spinal cord to the caudal end of the dura-arachnoid sac at S2.

8.030 All but one of the following statements regarding the arachnoidal villi are correct:
A. Permit one-way flow of CSF through the endothelium of dural venous sinuses.
B. Project into dural venous sinuses.
C. Are lined by pia mater.
D. Play a role in the transfer of CSF to venous blood.
E. May become calcified and visible on x-ray.

C is the correct answer.
The arachnoid villi are lined by arachnoid as they protrude into the dural venous sinuses. The pia stays behind, intimately attached to the CNS substance. CSF is passed through the villi into dural sinus venous blood only when CSF pressure exceeds dural sinus venous pressure. There is no fluid movement allowed from dural sinus to subarachnoid space. Arachnoid granulations often become calcified, and thus visible on x-ray.

SECTION 9: EMBRYOLOGY

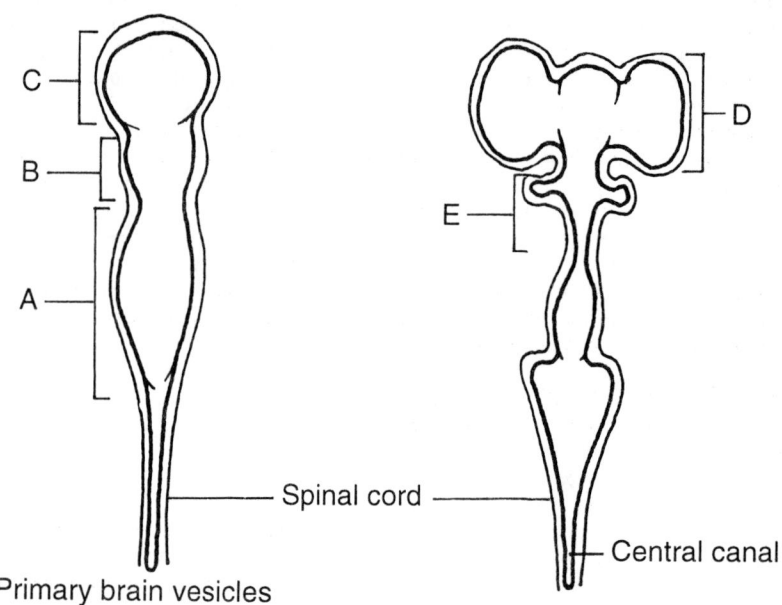

Primary brain vesicles

For questions 1 - 4, refer to the diagrams above.

9.001 In the above diagrams, _____ grows and differentiates to form the cerebral hemispheres.

D is the correct answer.

9.002 In the above diagrams, _____ becomes the prosencephalon.

C is the correct answer.

9.003 In the above diagrams, _____ develops to form the pons and medulla.

A is the correct answer.

9.004 In the above diagrams, _____ becomes the diencephalon.

E is the correct answer.

9.005 Which of the following would lie lateral to the sulcus limitans (hence, alar plate)?
A. Hypoglossal.
B. Trochlear.
C. Facial
D. Chief sensory nucleus of V

D is the correct answer.
Alar plate contains sensory neurons. The rhombencephalon opens like a book in its dorsal aspect. This shifts the alar plate from a dorsal position to a lateral position in relation to the sulcus limitans.

9.006 The cranial end of the neural tube develops three enlargements termed the prosencephalon, mesencephalon, and rhombencephalon. Which one of the following statements is incorrect?
A. Each enlargement surrounds and is derived from a neuroepithelium-lined primary brain vesicle.
B. The prosencephalon develops into two segments - the diencephalon and the telencephalon.
C. The mesencephalon does not divide, and forms the adult midbrain.
D. The cerebellum develops as an outgrowth of the diencephalon.
E. The primary brain vesicle remaining in the adult midbrain is called the Aqueduct of Sylvius.

D is the correct answer.
The rhombencephalon develops into two segments: metencephalon and myelencephalon. Metencephalon produces the pons and the cerebellum; the myelencephalon is the medulla. The entire brain is developed from the three primordial brain vesicles.

9.007 In the fetus which one of the following structures develops from the myelencephalon?
A. Superior olivary nucleus.
B. Hypoglossal nucleus.
C. Dentate nucleus.
D. Facial colliculus.
E. Inferior colliculus.

B is the correct answer.
The hypoglossal is the 12th and last and lowest of the cranial nerves excluding the spinal portion of XI. It is medullary (myelencephalon) all the way. All the other items develop from more cranial segments.

9.008 Identify the correct statement.
A. The telencephalon develops into the diencephalon and mesencephalon.
B. The lamina terminalis marks the cranial end of the embryonic neural tube.
C. The corpus callosum, anterior commissure, posterior commissure are all fiber tracts enabling communication between telencephalic structures.
D. The corpora quadrigemina constitute the midbrain tegmentum.
E. The prosencephalon forms the diencephalon and telencephalon which are joined by the cerebral aqueduct.

B is the correct answer.
As the name suggests, the terminal (most rostral) part of the three primitive brain vesicles becomes the lamina terminalis in the formed brain. The telencephalon does not "divide". Posterior commissure is a communicator of the diencephalon. The corpora quadrigemina form the tectum of the midbrain, the mesencephalon.

9.009 Identify the correct statement.
A. The significance of the sulcus limitans is that it divides the gray matter from the white matter in the spinal cord.
B. Neuroepithelium forming the lining of the embryonic neural tube forms all the neurons of the central nervous system.
C. The embryonic marginal layer becomes the gray matter of the cord and nuclear regions of the brainstem.
D. The alar plate of the embryonic cord will contain the motor neurons of the developed cord.
E. The sulcus limitans is not visible in the adult brainstem.

B is the correct answer.
Neurons outside the central nervous system come from the neural crest. The sulcus limitans separates the alar (sensory) and basal (motor) plates of developing gray matter. It is visible throughout the fourth ventricle floor. The marginal layer consists of myelinated processes of neurons in the gray matter. The myelin gives a paler look to the marginal layer - thus, white matter.

9.010 Which one of the following pairs demonstrates a developmental relationship?
A. Notochord --- Nucleus pulposus.
B. Neural tube --- Subarachnoid space.
C. Neural crest --- Nucleus ambiguus.
D. Alar plate --- GVE neurons.
E. Prosencephalon --- Aqueduct of Sylvius.

A is the correct answer.
The neural tube has no developmental responsibility for mesodermal tissues such as the meninges. The neural crest does not supply CNS neurons. The alar plate is involved with sensory, not motor activity. The Aqueduct is a remnant of the mesencephalic vesicle.

9.011 Derivatives of the neural crest include all but one of the following:
A. Dorsal root ganglion cells.
B. Cells of the geniculate nucleus of the facial nerve.
C. Celiac ganglion neurons.
D. Cells of the nucleus of the spinal tract of the trigeminal nerve.
E. Cells of the adrenal medulla.

D is the correct answer.
The neural crest is the source of all neurons outside the central nervous system including cells of the adrenal medulla. Thus the cells of the spinal tract nucleus are from the alar plate neuroepithelium, not from the neural crest.

9.012 Which one of the following statements is INCORRECT?
A. The neuroepithelium lining the early neural tube gives rise to neuroblasts and glioblasts.
B. The tissue lining the neural tube, the ependyma, is of endodermal origin.
C. Neuroblasts derived from neural crest tissue give rise to all the primary sensory neurons of the peripheral nervous system and all postganglionic neurons.
D. The central cavity of the early neural tube persists in the adult as the central canal of the spinal cord.
E. Failure of proper closure of the neural folds leads to varying degrees and locations of spina bifida.

B is the correct answer.
Nervous tissue is ectodermal in origin, stemming from infolding of dorsal embryonic ectoderm to form a neural tube. Failure of complete closure of the infolding in either the cranial or caudal neuropore areas results in spina bifida in these locations. The central canal of the spinal cord is indeed the persistence of the primordial neural tube cavity.

9.013 Which of the following structures does not develop from the neural crest?
A. Dorsal root ganglia.
B. Adrenal medulla.
C. Parasympathetic ganglia.
D. Sympathetic preganglionic neurons.
E. Sympathetic trunk neurons.

D is the correct answer.
The neurons outside of the central nervous system are developed from the neural crest. This would include all but "D". The CNS itself - brain and spinal cord - develops from the neural tube. This would include the preganglionic neurons of the spinal cord lateral horns.

9.014 Choose the CORRECT association:
A. Basal plate - general somatic efferent.
B. Dorsal roots - myotome.
C. Diencephalon - cerebral aqueduct (of Sylvius).
D. Cisterna magna - subdural space.
E. Prosencephalon - midbrain.

A is the correct answer.
The basal plate is associated with motor neuron development, including GSE neurons. Correct pairings for the others might be: Dorsal roots - Dermatomes; Diencephalon - Third ventricle; Cisterna magna - Subarachnoid space; Prosencephalon - Diencephalon or telencephalon

9.015 Select the INCORRECT statement regarding the neural crest:
A. It is derived from primitive embryonic ectoderm.
B. In addition to the obvious neurons of the peripheral nervous system, it gives rise to cells of the adrenal medulla, melanocytes, and contributes to the formation of the pia mater.
C. It forms the neurilemma cells, thus the covering of peripheral nerves.
D. If not properly formed, results in the condition of spina bifida.
E. Arises from tissue of the embryonic neural fold.

D is the correct answer.
Neural crest derivatives include cells forming sheaths of Schwann, melanocytes and pia, as well as neurons as above. Spina bifida is not the fault of the neural crest but of faulty closure of the neural folds. The crest does arise from tissue in the area of the neural folds, but is separated before closure of the folds.

9.016 Which one of the following statements is INCORRECT?
A. Cranial nerve neurons derived from the basal plate of the mantle layer of the embryonic cord are motor in function.
B. Sympathetic GVE cells arise from the mantle layer just ventral to the sulcus limitans.
C. The sulcus limitans demarcates the division between the alar and basal plates.
D. The alar plate gives rise to neurons whose fibers form the ventral roots of the spinal nerves.
E. The cranial and caudal neuropores of the neural tube are the last areas of the tube to close.

D is the correct answer.
Ventral root fibers are processes of neurons residing in the basal plate. The sulcus limitans demarcates the division between the motor basal and the sensory alar plates. GVE neurons, as exemplified by the lateral horns of T1 - L2 are indeed located just ventral to the sulcus limitans. Areas of terminal neural tube closure are at the anterior and posterior neuropores

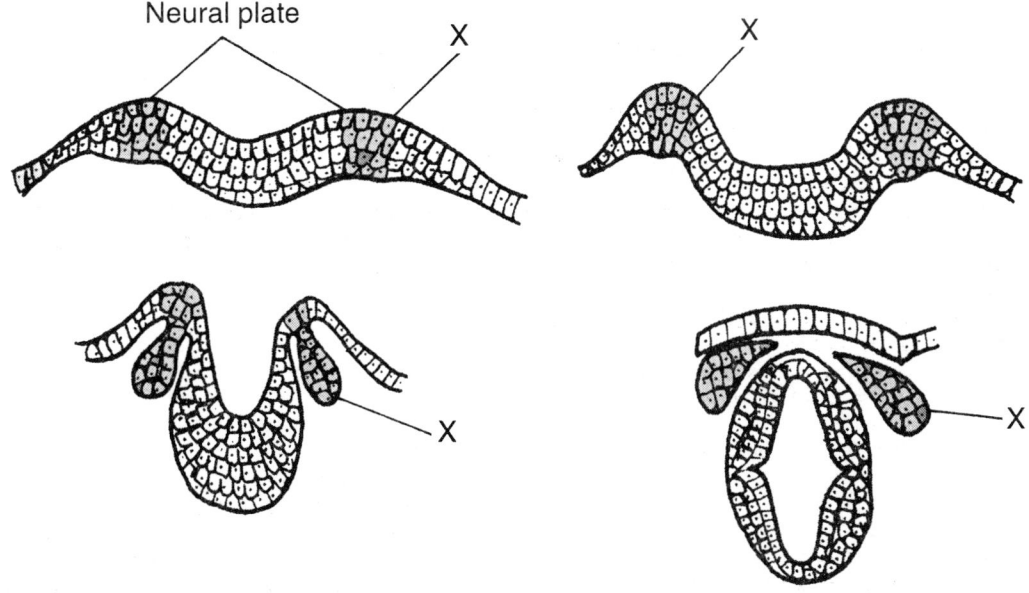

For question 17, refer to the diagrams above.

9.017 In the drawings above, "X" represents:
A. Neural crest.
B. Mesoderm.
C. Precursor of a vertebral body.
D. Ligamentum flavum.
E. Precursor of meningeal structures.

A is the correct answer.
Neural crest tissue grows and separates from the ectoderm of the neural plate. At an early stage it is seen perched dorsolaterally above the neural tube. The neural crest has to do with peripheral nervous tissue; not vertebral bodies or ligaments, nor mesodermal meninges.

SECTION 10: EYE & EAR

10.001 Which of the following statements regarding the visual pathway is false?
A. Fibers from the nasal half of each retina cross in the optic chiasm.
B. The right optic tract is composed of fibers of ganglion cells located in the left nasal retina and the right temporal retina.
C. The brachium of the superior colliculus is composed of retinal fibers to the preoptic nucleus and the superior colliculus.
D. All retinal fibers synapse in the lateral geniculate body.
E. Projection from the lateral geniculate is to the ipsilateral calcarine cortex.

D is the correct answer.
A number of retinal fibers continue without synapse past the lateral geniculate and, via the brachium of the superior colliculus, continue to the superior colliculus and the pretectal area. These fibers are responsible for visual and head-eye reflex performance via the teltospinal tract. Nasal retinal fibers cross in the chiasm; temporal fibers do not. Thus the right optic tract is indeed formed from right temporal and left nasal retinal nerve fibers.

10.002 The occipital primary visual cortex:
A. is embryologically a diverticulum of the diencephalon.
B. is arranged so that the opposite lower visual field is represented above the calcarine fissure (upper bank)
C. is a primitive cortex with only a single layer of pyramidal cells, similar to the hippocampus.
D. has a macular field representation most anteriorly in the primary visual cortical area.
E. projects to visual association cortex surrounding the lateral geniculate body.

B is the correct answer.
The primary visual cortex is a complex cortex arranged as described in answer "B". Macular vision is represented in its most posterior part. Surrounding the primary visual cortex is the visual association cortex (areas 18 and 19 Brodmann). These are a far piece from the lateral geniculate! The eye itself is a diverticulum of the diencephalon - note the termination of the visual retinal fibers in the diencephalic lateral geniculate body.

10.003 Transection of the left oculomotor nerve results in all but one of the following:
A. Dilation of the left pupil.
B. Abduction of the right eye.
C. Loss of light reflex left.
D. Diplopia.
E. No change in the corneal reflex.

B is the correct answer.
Loss of left cranial nerve III results in an abducted position of the left eye and diplopia, due to unopposed pull of the abducens innervated lateral rectus muscle. The left pupil will dilate due to unopposed action of the sympathetic innervated iris dilator. The left pupil will be unable to constrict in response to light due to loss of function of the iris constrictor innervated by parasympathetics carried in the oculomotor nerve. The corneal reflex involves cranial nerves V and VII.

10.004 Diplopia (double vision):
A. is typically caused by lesions of the calcarine cortex.
B. has no relationship to nystagmus.
C. is usually the result of extraocular muscle imbalance.
D. is frequently the result of retinal lesions.
E. is not relieved by covering one eye with a patch.

C is the correct answer.
Diplopia occurs when the eyes are not working in congruent fashion. The usual cause is extraocular muscle imbalance for whatever reason. Covering one eye will obviously relieve the double vision. Both eyes are required to produce two disparate images. Cortical or retinal defects produce field changes, not double vision.

10.005 Horner's syndrome: All the following statements are correct EXCEPT:
   A. Is seen in spinal cord lesions above T1.
   B. Is ipsilateral to the lesion.
   C. Is a prominent feature of the findings secondary to local damage of the dorsal motor nucleus of the vagus.
   D. Examination reveals ipsilateral pupillary constriction and slight palpebral ptosis.
   E. May result from lesions involving descending autonomic fibers in the brainstem.

C is the correct answer.
The path of sympathetic fibers to the eye includes reticular tracts in the brainstem tegmentum, lateral horn cells of T1, motor root of T1, white ramus communicans of T1, cervical sympathetic trunk, superior cervical ganglion, and the carotid nerve. Interruption at any point will result in an ipsilateral Horner's syndrome. The pathway does not cross. The vagus has nothing to do with Horner's syndrome.

10.006 The oculomotor nerve:
   A. arises in the upper pons rostral to the crossing of the brachium conjunctivum.
   B. innervates branchiogenic muscles of the orbit.
   C. distributes to all the extraocular muscles except the superior oblique and the lateral rectus.
   D. travels with the optic nerve through the optic canal to reach the orbit.
   E. includes sympathetic fibers which join it in the interpeduncular fossa.

C is the correct answer.
Oculomotor nerve arises from a nucleus in the midbrain and runs a course, entirely separate from the optic nerve, through the cavernous sinus and into the orbit via the superior orbital fissure. It does not traverse the optic canal. The oculomotor innervates somitic muscles as listed in answer "C". Thus its component is GSE. A few sympathetic fibers communicate with the oculomotor nerve in the cavernous sinus, taking off from the sympathetic plexus about the internal carotid artery. These are sympathetic fibers that run through the ciliary ganglion without synapse to reach vessels of the bulb and ciliary and iris musculature. The bulk of the sympathetics enter the orbit by riding on the friendly nasociliary nerve, eventually leaving this structure as the long ciliary nerves.

10.007 The eye, with complete loss of III nerve function, is:
   A. pulled medially by residual medial rectus function.
   B. pulled laterally by residual lateral rectus function.
   C. pulled medially and downward by residual medial rectus and superior oblique function.
   D. not changed in position.
   E. is afflicted with nystagmus.

B is the correct answer.
Unopposed pull of the abducens nerve produces this deviation. Deviation by the unopposed superior oblique muscle is more rotary intortion than visible deviation in this abducted position. Nystagmus is not stimulated by oculomotor dysfunction.

10.008 A lesion of one medial longitudinal fasciculus (MLF):
   A. causes weakness of the ipsilateral medial rectus muscle for both horizontal gaze and convergence.
   B. causes dysfunction of the Edinger-Westphal nucleus with resulting diplopia.
   C. only causes weakness of convergence.
   D. causes weakness of the ipsilateral medial rectus muscle in horizontal gaze, but not convergence.
   E. does not cause any disability in vision or extra-ocular muscle function.

D is the correct answer.
The right and left medial longitudinal fasciculi tie together the medial and lateral rectus muscles through connections with the paramedian pontine reticular formation (PPRF) and the oculomotor nuclei. This connection is essential for the proper conjugate function of the eyes in right and left horizontal gaze. If one eye abducts, the other must adduct to an equal degree, as stimulated by fibers in the ipsilateral MLF. This path is not utilized in convergence nor in autonomic eye function. If one MLF is injured the ipsilateral eye will not adduct on attempts at horizontal gaze to the opposite side. Convergence adduction will remain intact.

10.009 Vertigo:
- A. is the illusion of self (subject) or environment (object) moving.
- B. is typically associated with double vision.
- C. cannot be induced by caloric water irrigation of the external auditory canal.
- D. when caused by spinning, cannot be suppressed.
- E. goes away when the subject's eyes are closed.

A is the correct answer.
This statement accurately describes vertigo. Double vision is not present. Semicircular canal caloric stimulation can produce eye movement, nystagmus and vertigo. Closing the eyes does not eliminate the illusion. Twirling figure skaters learn how to suppress vertiginous stimuli – otherwise they would have a dizzyingly hard time!

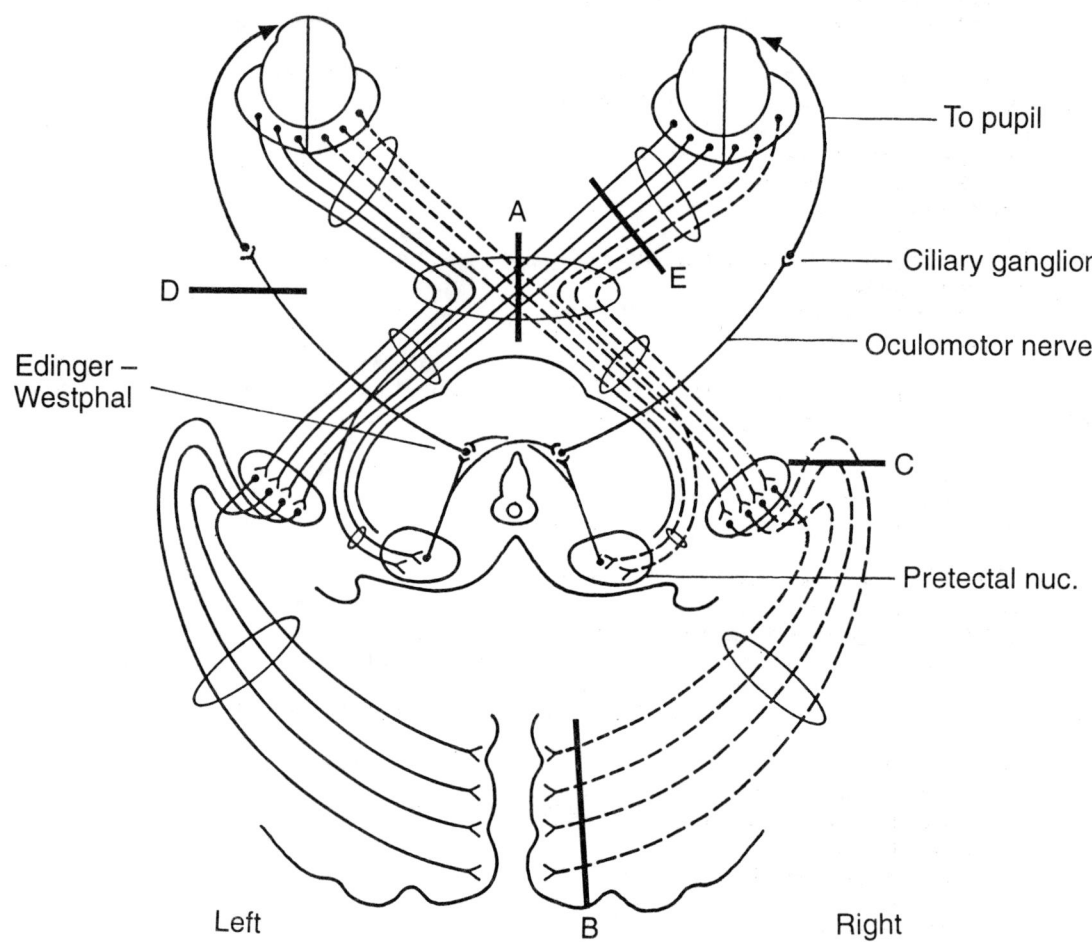

For questions 10 - 13, refer to the diagram above.

10.010 In the above diagram, a lesion at location _____ would produce homonymous hemianopia.

B is the correct answer.

10.011 In the above diagram, a lesion at location _____ would abolish direct and consensual pupillary light reflexes.

D is the correct answer.

10.012 In the above diagram, a lesion at location _____ would result in the pupil on this side would not react directly to light, but the other pupil would react consensually.

D is the correct answer.

10.013 In the above diagram, a lesion at location _____ would produce loss of vision in the upper left visual field - a quadrantic homonymous hemianopia.

C is the correct answer.

10.014 With loss of the right optic nerve, which of the following would be true during resting?
   A. Both pupils normal resting size
   B. Both pupils smaller than resting size
   C. Right pupil smaller than resting size, Left pupil normal resting size
   D. Right pupil larger than resting size, Left pupil normal resting size
   E. Right pupil larger than resting size, Left pupil smaller than resting size

A is the correct answer.

10.015 With loss of the right optic nerve, which of the following would be true concerning the reaction to light in the right eye?
   A. Both pupils normal resting size
   B. Both pupils smaller than resting size
   C. Right pupil smaller than resting size, Left pupil normal resting size
   D. Right pupil larger than resting size, Left pupil normal resting size
   E. Right pupil larger than resting size, Left pupil smaller than resting size

A is the correct answer.

10.016 With loss of the right optic nerve, which of the following would be true concerning the reaction to light in the left eye?
   A. Both pupils normal resting size
   B. Both pupils smaller than resting size
   C. Right pupil smaller than resting size, Left pupil normal resting size
   D. Right pupil larger than resting size, Left pupil normal resting size
   E. Right pupil larger than resting size, Left pupil smaller than resting size

B is the correct answer.

10.017 With loss of the right cervical sympathetic trunk, which of the following would be true during resting?
   A. Both pupils normal resting size
   B. Both pupils smaller than resting size
   C. Right pupil smaller than resting size, Left pupil normal resting size
   D. Right pupil larger than resting size, Left pupil normal resting size
   E. Right pupil larger than resting size, Left pupil smaller than resting size

C is the correct answer.

10.018 With loss of the right cervical sympathetic trunk, which of the following would be true concerning the reaction to light in the right eye?
   A. Both pupils normal resting size
   B. Both pupils smaller than resting size
   C. Right pupil smaller than resting size, Left pupil normal resting size
   D. Right pupil larger than resting size, Left pupil normal resting size
   E. Right pupil larger than resting size, Left pupil smaller than resting size

B is the correct answer.

10.019 With loss of the right cervical sympathetic trunk, which of the following would be true concerning the reaction to light in the left eye?
- A. Both pupils normal resting size
- B. Both pupils smaller than resting size
- C. Right pupil smaller than resting size, Left pupil normal resting size
- D. Right pupil larger than resting size, Left pupil normal resting size
- E. Right pupil larger than resting size, Left pupil smaller than resting size

B is the correct answer.

10.020 With loss of the right oculomotor nerve, which of the following would be true during resting?
- A. Both pupils normal resting size
- B. Both pupils smaller than resting size
- C. Right pupil smaller than resting size, Left pupil normal resting size
- D. Right pupil larger than resting size, Left pupil normal resting size
- E. Right pupil larger than resting size, Left pupil smaller than resting size

D is the correct answer.

10.021 With loss of the right oculomotor nerve, which of the following would be true concerning the reaction to light in the right eye?
- A. Both pupils normal resting size
- B. Both pupils smaller than resting size
- C. Right pupil smaller than resting size, Left pupil normal resting size
- D. Right pupil larger than resting size, Left pupil normal resting size
- E. Right pupil larger than resting size, Left pupil smaller than resting size

E is the correct answer.

10.022 With loss of the right oculomotor nerve, which of the following would be true concerning the reaction to light in the left eye?
- A. Both pupils normal resting size
- B. Both pupils smaller than resting size
- C. Right pupil smaller than resting size, Left pupil normal resting size
- D. Right pupil larger than resting size, Left pupil normal resting size
- E. Right pupil larger than resting size, Left pupil smaller than resting size

E is the correct answer.

10.023 Regarding the pupillary light reflex, which one of the following statements is INCORRECT?
- A. This is an afferent-efferent system with the afferent arm the retinal system, and the oculomotor nerve the efferent arm.
- B. Is inoperative in cases of bilateral destruction of the occipital lobes.
- C. Is a direct (ipsilateral) and consensual (contralateral) pupillary constrictor response to the exhibition of light.
- D. Is mediated by efferent preganglionic parasympathetic fibers which arise in the Edinger-Westphal nucleus.
- E. The posterior commissure in the midbrain pretectum helps insure a consensual response.

B is the correct answer.
The occipital lobes and their visual cortices are not involved in the production of the pupillary light reflex. Retinal fibers carry the light stimulus to the pretectal nuclei which are bilaterally informed (posterior commissure). Directions flow from the pretectal nuclei to the Edinger-Westphal nucleus, then out the III nerve to the ciliary ganglion, and then the pupil.

10.024 The visual fields:
  A. show homonymous defects with lesions of the lateral geniculate body.
  B. are represented centrally by rods of the retina; laterally by the cones.
  C. can only be tested with an ophthalmoscope.
  D. tend to be exclusive to each eye with no binocular overlap.
  E. Depend on the brachium of the superior colliculus for accuracy when tested.

A is the correct answer.
Binocular fields are characteristic of visual function. They may be tested easily at the bedside by confrontational comparison with the examiner's normal fields. Retinal rods pick up peripheral fields, cones are clustered in the center. By the time the retinal fibers reach the lateral geniculate body, they have been sorted at the chiasm so that the right world is delivered to the left lateral geniculate, and vice versa. Lateral geniculate lesions therefore produce homonymous defects (i.e., the right or left halves of the visual fields of both eyes). The brachium of the superior colliculus is not involved in field testing. It carries fibers for reflex activity, not vision per se.

10.025 With destruction of the right optic nerve:
  A. A light shined in the left eye will produce constriction of the left pupil only.
  B. A light shined in the right eye will produce constriction of both pupils.
  C. A light shined in the right eye will produce only constriction of the left pupil.
  D. A light shined in the left eye will produce constriction in both pupils.
  E. The corneal reflex on the right will be abolished.

D is the correct answer.
With a dead optic nerve, no visual stimulus can be appreciated, no light can be perceived, no visual reflex can be induced in that eye. However, if nothing is wrong with the oculomotor nerve on the side of the dead optic nerve, consensual activity will still be present on the dead side. The corneal reflex involves trigeminal and facial nerves; the optic nerve is not involved.

10.026 Nystagmus is:
  A. the result of lesions in cranial nerves III and VI.
  B. a common finding in figure skaters after doing a spin.
  C. a combination of fast (reflex) and slow (checking) movement of the eyes.
  D. a combination of fast (checking) and slow (reflex) movement of the eyes.
  E. cannot be induced in cases of conduction deafness.

D is the correct answer.
Nystagmus requires intact III and VI nerves. The eye movement of nystagmus is a slow tracking reflex movement followed by a rapid correcting check. Figure skaters learn to inhibit the reflex movement through cortical activity - otherwise they would not know whether they were coming or going! Nystagmus may be induced by caloric testing as long as the semicircular canals and the VIII nerve are intact.

10.027 Horizontal gaze:
  A. depends on the integrity of the paramedian pontine reticular formation (PPRF).
  B. is mostly influenced by the deep cerebellar nuclei.
  C. is predominantly a reflex with little voluntary contribution.
  D. becomes defective as far as muscular action is concerned when the optic nerve in one eye is destroyed.
  E. is defective as far as muscular action is concerned following lesions of the ciliary ganglion.

A is the correct answer.
Horizontal gaze is obviously a voluntary act. The eyes are yoked together by the activity of the paramedian pontine reticular formation (PPRF) about the abducens nucleus. Neither the cerebellum, optic nerve, nor the ciliary ganglion are required.

10.028 Select the one INCORRECT statement. Inability of the right eye to adduct when attempting lateral gaze to the left may be due to:
  A. right oculomotor nerve paralysis.
  B. lesion of the MLF on the right.
  C. destruction of the PPRF on the left.
  D. right medial rectus injury.
  E. loss of vision in the right eye.

E is the correct answer.
Inability to adduct the eye may be due to a defective oculomotor nerve, right medial rectus muscle injury, or a lesion of the PPRF/MLF yoking structure. In this case, adduction of the right eye depends on the left PPRF and the right MLF. Actual vision has nothing to do with the adduction motion.

10.029 Which of the following is not an integral part of the visual accommodation pathway?
A. PPRF.
B. Pretectal nucleus of the midbrain.
C. Ciliary ganglion.
D. Edinger-Westphal nucleus.
E. Visual cortex.

A is the correct answer.
Accommodation obviously requires the ability to see, and integrity of communication between calcarine cortex and the pretectal area. Granted all that, local requirements involve an intact vergence center, oculomotor nerves, and medial rectus muscles. Also the Edinger-Westphal-ciliary ganglion parasympathetic reflex apparatus must be working. Convergence, pupillary and lens focusing changes result. The PPRF horizontal gaze center is not involved.

10.030 Inferior oblique muscle is most closely associated with:
A. Edinger-Westphal nucleus.
B. Abducens nerve.
C. GSE oculomotor nucleus.
D. Trochlear nerve.
E. Ventral root T1.

C is the correct answer.

10.031 Iris dilator (radial muscle) is most closely associated with:
A. Edinger-Westphal nucleus.
B. Abducens nerve.
C. GSE oculomotor nucleus.
D. Trochlear nerve.
E. Ventral root T1.

E is the correct answer.
The tortuous route of sympathetics to the eye involves stimuli carried in reticular system fibers through the brainstem tegmentum and down the cervical cord to the lateral horn neurons of T1. Axons of these neurons exit the cord in the T1 ventral root.

10.032 Superior oblique muscle is most closely associated with:
A. Edinger-Westphal nucleus.
B. Abducens nerve.
C. GSE oculomotor nucleus.
D. Trochlear nerve.
E. Ventral root T1.

D is the correct answer.

10.033 Lens diameter is most closely associated with:
A. Edinger-Westphal nucleus.
B. Abducens nerve.
C. GSE oculomotor nucleus.
D. Trochlear nerve.
E. Ventral root T1.

A is the correct answer.
Lens diameter is a parasympathetic responsibility. Preganglionic fibers originate in the Edinger-Westphal nucleus and are carried to the ciliary ganglion in the oculomotor nerve. Postganglionics proceed to the ciliary body smooth musculature (ciliaris muscle). Contraction of the circular ciliaris muscle draws in the ciliary processes which anchor the suspensory ligaments of the lens. The resulting smaller circle allows relaxation of the pull of the lens ligaments and thus permits rounding up of the lens itself - a form it prefers, but is controlled by the tension of the lens suspensory ligaments.

10.034 Iris constrictor (sphincter muscle) is most closely associated with:
A. Edinger-Westphal nucleus.
B. Abducens nerve.
C. GSE oculomotor nucleus.
D. Trochlear nerve.
E. Ventral root T1.

A is the correct answer.
Iris constriction is similarly a parasympathetic process (the sympathetics simultaneously inhibit the dilator muscle fibers). The pathway is the same as in the previous description except that the postganglionics from the ciliary ganglion extend to the smooth muscle circular fibers of the sphincter pupillae.

10.035 Convergence is most closely associated with:
A. Oculomotor nerve.
B. Cervical sympathetic trunk.
C. Optic nerve.
D. Trochlear nerve.
E. Visual cortex.

A is the correct answer.
Convergence is operated by the oculomotor nerve under direction of a vergence center in the midbrain.

10.036 Pupillary dilation is most closely associated with:
A. Oculomotor nerve.
B. Cervical sympathetic trunk.
C. Optic nerve.
D. Trochlear nerve.
E. Visual cortex.

B is the correct answer.
Sympathetics enter the head primarily through postganglionics generated in the superior cervical ganglion of the sympathetic trunk. These fibers ride on the internal carotid artery. The postganglionic fibers from the superior cervical ganglion to the internal carotid are termed the "carotid nerve".

10.037 Initiation of accommodation is most closely associated with:
A. Oculomotor nerve.
B. Cervical sympathetic trunk.
C. Optic nerve.
D. Trochlear nerve.
E. Visual cortex.

E is the correct answer.

10.038 Loss of efferent limb of the light reflex is most closely associated with:
A. Oculomotor nerve.
B. Cervical sympathetic trunk.
C. Optic nerve.
D. Trochlear nerve.
E. Visual cortex.

A is the correct answer.
Efferents of the light reflex are parasympathetics carried in the oculomotor nerve.

10.039 Right homonymous hemianopia is most closely associated with:
A. Macula of the retina.
B. Optic chiasm.
C. Peripheral retinal receptors.
D. Left calcarine cortex.
E. right parietal lobe.

D is the correct answer.
To produce homonymous loss of half of the visual fields, interruption of all optic fibers, after sorting in the chiasm, must occur. This may involve the optic tract, lateral geniculate body, optic radiation, or the calcarine cortex itself. The result is the same.

10.040 Left inferior homonymous quadrantic hemianopia is most closely associated with:
A. Macula of the retina.
B. Optic chiasm.
C. Peripheral retinal receptors.
D. Left calcarine cortex.
E. right parietal lobe.

E is the correct answer.
Optic radiations stream back to the calcarine cortex in two main portions - either directly back through the parietal lobe, or looping around the temporal horn of the lateral ventricle in the temporal lobe (Meyer's loop). The latter fibers reach the calcarine cortex by running through the temporal lobe close to the lateral aspect of the lateral ventricle. Parietal lobe radiation carries fibers representing the inferior visual fields (superior retinas). Loss of the parietal fibers will result in loss of one half of the optic radiation on one side, or a homonymous quadrantic defect found in the inferior fields. Of course, a similar defect will be produced by a lesion of the inferior bank of the calcarine cortex itself.

10.041 Central blindness (central scotoma) is most closely associated with:
  A. Macula of the retina.
  B. Optic chiasm.
  C. Peripheral retinal receptors.
  D. Left calcarine cortex.
  E. right parietal lobe.

A is the correct answer.

10.042 Night blindness is most closely associated with:
  A. Macula of the retina.
  B. Optic chiasm.
  C. Peripheral retinal receptors.
  D. Left calcarine cortex.
  E. right parietal lobe.

C is the correct answer.
Peripheral retinal rods react to a low intensity of light. They perceive objects better than the central cones in twilight or night.

10.043 Bitemporal hemianopia is most closely associated with:
  A. Macula of the retina.
  B. Optic chiasm.
  C. Peripheral retinal receptors.
  D. Left calcarine cortex.
  E. right parietal lobe.

B is the correct answer.
Nasal retinal fibers cross in the optic chiasm. They represent the temporal visual fields. Injury of the chiasm will then result in blindness in these temporal fields bilaterally.

10.044 The left visual field is represented in the:
  A. right lateral geniculate nucleus.
  B. left lateral geniculate nucleus.
  C. postcentral gyrus right.
  D. visual cortex of the left hemisphere.
  E. left superior colliculus.

A is the correct answer.

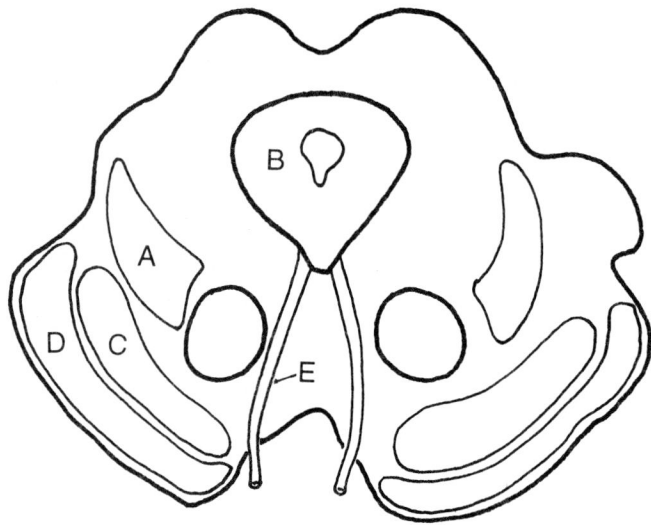

For question 45, refer to the diagram above.

10.045 Damage to which of the labeled (A-E) structures above will interfere with the pupillary light reflex?

E is the correct answer.
This is the oculomotor nerve which carries parasympathetic fibers from the Edinger-Westphal nucleus to the ciliary ganglion; thence to the iris muscles. A. is the medial lemniscus, B. the periaqueductal gray matter, C. the substantia nigra, and D. the cerebral peduncle. None of these is concerned with the pupillary light reflex.

10.046 Intorsion of the eye is:
A. the regression of the eye back into the orbit.
B. the inward rotation of the upper portion of the eye.
C. the downward movement (depression) of the eye.
D. the inward or adduction movement of the eye.
E. None of the above.

B is the correct answer.
Intorsion = "twisting inward". This motion of the eye is produced by the superior oblique eye muscle. Regression of the eye is caused by dysfunction of the smooth muscle of the orbit (enophthalamus of Horner's syndrome). Downward movement is caused by the combined action of the superior oblique and the inferior rectus. Adduction is the province of the medial rectus muscle.

10.047 Interruption of which pathway in the pons will eliminate conjugate horizontal eye movements?
A. Medial lemniscus.
B. Lateral lemniscus.
C. Central tegmental tract.
D. Inferior cerebellar peduncle.
E. Medial longitudinal fasciculus.

E is the correct answer.
The horizontal motion center lies in the paramedian pontine reticular formation (PPRF). Communication between this "yoking" center and the medial and lateral recti of the eye is carried in the medial longitudinal fasciculus. The MLF must be intact to perform conjugate horizontal eye movements. The other tracts are not involved.

10.048 At what point in the visual system is input from both eyes first "sorted" so that information from the left visual field is all located in the right hemisphere?
A. Optic nerve.
B. Optic chiasm.
C. Lateral geniculate body.
D. Primary visual cortex.
E. Corpus callosum.

B is the correct answer.
This "arranging" function is the raison d'etre of the optic chiasm (crossing). Fibers from similar areas of each retina are brought together. The optic nerves carry the fibers that must be rearranged. The lateral geniculate and primary visual cortex receive the rearranged fibers. The corpus callosum plays no role in this regard.

10.049 In looking from a distant to a near object, the pupil:
A. dilates due to increased sympathetic activity.
B. constricts due to increased sympathetic activity.
C. dilates due to decreased sympathetic activity.
D. constricts due to increased oculomotor nerve parasympathetic activity.
E. does not change in diameter because the sympathetic and oculomotor effects counteract each other.

D is the correct answer.
For sharper definition of a near object, the pupil contracts due to parasympathetic activity - fibers are in the III nerve. Sympathetic activity dilates the pupil. We are good enough to actively stimulate or inhibit opposing neuromuscular action in all areas!

10.050 Which nerve is not functioning properly if the pupillary light reflex is not seen in either eye after illuminating the left eye, but is seen in both eyes after illuminating the right eye?
A. Left optic nerve.
B. Right optic nerve.
C. Left oculomotor nerve.
D. Right oculomotor nerve.
E. Left abducens nerve.

A is the correct answer.
The sensory arm of the pupillary light reflex is the retina. The retinal message is carried in the optic nerve. The motor arm for pupil constriction is carried in the oculomotor nerve parasympathetic fibers from the Edinger-Westphal nucleus. Thus, when the left optic nerve is knocked out, the (visual) sensory function of the left eye is zero. However, as the right is O.K. and the motor arm of the light reflex is carried in the III nerve, both pupils will constrict when light is shined into the good eye.

For question 51, refer to the diagrams below.

10.051 Match the numbered optic pathway lesion to the lettered visual field defect. (One lettered visual field is used twice.)

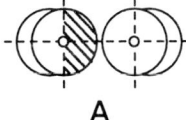
A

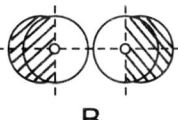
B

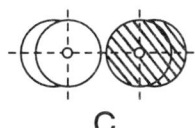
C

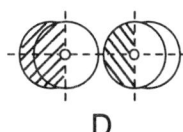
D

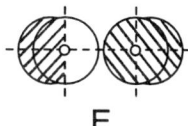
E

10.052 A cause of nerve deafness is:
   A. perforated tympanic membrane.
   B. middle ear infection.
   C. arthritis of the middle ear ossicles.
   D. prolonged exposure to loud sounds
   E. impacted cerumen (ear wax).

D is the correct answer.
Prolonged exposure to loud sounds destroys the delicate hair cell mechanism of the Organ of Corti. The deafness of soldiers after prolonged exposure to the roar of battle is a good example. The other statements all refer to conduction difficulties. Middle ear infections scar the delicate intraossicular joints and scar the tympanum.

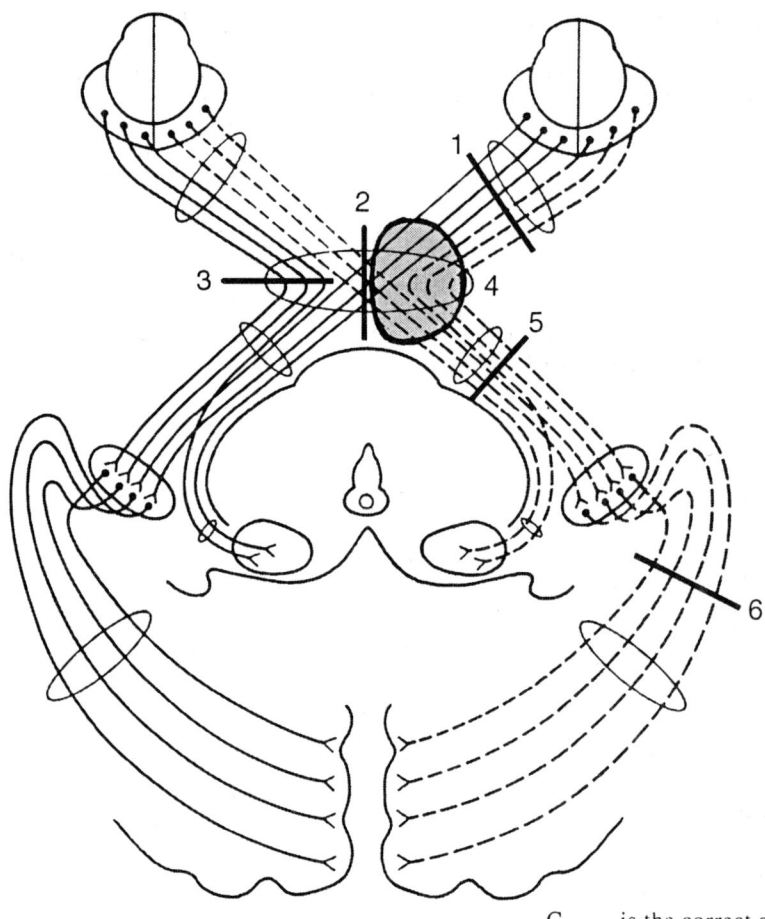

Lesion 1		C	is the correct answer
Lesion 2	(midline severing the optic chiasm)	B	is the correct answer
Lesion 3	(tumor eroding the lateral edge of the chiasm)	A	is the correct answer
Lesion 4	(pituitary tumor indicated)	E	is the correct answer
Lesion 5		D	is the correct answer
Lesion 6		D	is the correct answer

10.053 Regarding cold water irrigation of the left ear, one statement below is INCORRECT:
   A. Nystagmus may be produced.
   B. The fast checking movement of nystagmus will be abolished with cerebral cortical destruction.
   C. With destruction of the right MLF adduction of the right eye will not occur despite a normal right oculomotor nerve.
   D. Produces a slow movement of the eyes to the left.
   E. With bilateral destruction of the MLF rostral to the PPRF, no eye movement will follow.

E is the correct answer.
Bilateral destruction of the MLF rostral to the PPRF destroys the yoking mechanism of conjugate horizontal gaze, and therefore, the adduction of one eye. Abduction will occur as long as the MLF can carry vestibular impulses to the abducens nuclei, and these nuclei are O.K. Abduction does not require the rostral part of the MLF in horizontal gaze; adduction does. The other statements are all correct. Fast checking motion of nystagmus is a cortical function. Cold irrigation left causes relative dominance of the right horizontal canal, with resultant slow movement of the eyes to the left.

SECTION 11: HYPOTHALAMUS

11.001 Which of the following statements is INCORRECT regarding the hypothalamus?
A. Is controlled by the pituitary gland.
B. Plays an important role in water balance.
C. Plays an important role in autonomic activity throughout the organism.
D. Weighs about four grams in the human adult.
E. Plays an important role in bodily temperature control.

A is the correct answer.
The hypothalamus controls the dispensing of anterior pituitary hormones by its production of "releasing" hormones. These permit the elaboration of anterior pituitary hormones into the bloodstream. The posterior pituitary is "stocked" with hypothalamic-produced hormones (antidiuretic hormone, oxytocin). ADH is vital in the control of water balance. Body temperature may be controlled by sympathetic induced vasomotion, sweating, shivering, pilarrection - all directed by temperature sensors in the hypothalamus. For all its multiple functions and nuclei, the adult hypothalamus displaces only about four grams.

11.002 One of the following statements is INCORRECT:
A. Anterior hypothalamic nuclei influence parasympathetic function.
B. Posterior hypothalamic nuclei influence sympathetic function.
C. Certain hypothalamic nuclei secrete hormone releasing factors.
D. The hypothalamic sulcus divides the dorsal from the ventral tiers of hypothalamic nuclei.
E. The hypothalamus might be termed the "head ganglion" of the autonomic system.

D is the correct answer.
The hypothalamic sulcus divides the hypothalamus from the dorsal thalamus. All the other statements are certainly true. Hypothalamic nuclei are not arranged in tiers as are those of the dorsal thalamus.

11.003 Select the one INCORRECT statement : The hypothalamus
A. controls the release of hormones from the anterior lobe of the pituitary gland.
B. controls visceral reactions to emotional and/or sensory stimuli.
C. has nothing to do with the limbic system.
D. is connected with the reticular system of the brainstem.
E. contains many nuclei, among which is the mammillary body.

C is the correct answer.
The hypothalamus is intimately connected with the limbic system and is tempered in its activity by this "newer" system and its supervisory connections. The remaining statements are correct. The hypothalamus, other than its endocrine responsibilities, functions through the arms of the reticular system.

11.004 Which of the following is not part of, or strongly connected with, the hypothalamus?
A. Mammillary bodies.
B. Fasciculus retroflexus.
C. Fornix.
D. Interpeduncular nucleus.
E. Pulvinar.

E is the correct answer.
The functions of the hypothalamus will proceed in good shape without the participation of the pulvinar. The fasciculus retroflexus and the interpeduncular nucleus are reticular system parts connected with the hypothalamus by the striae medullares thalami and the habenular nuclei. The retroflexus is the link between habenula and the midbrain reticular interpeduncular nucleus.

11.005 Find the one INCORRECT statement regarding the hypothalamus.
   A. Communicates with the dorsal nucleus of the vagus via the reticular system.
   B. Controls both the endocrine and autonomic systems.
   C. Its neurons can secrete hormones.
   D. Communicates with the cerebral cortex through the thalamic fasciculus.
   E. Receives input from the amygdala.

D is the correct answer.
The thalamic fasciculus carries the ansa lenticularis, the lenticular fasciculus and the dentatothalamic tract to the VA/VL thalamic nuclei. The hypothalamus communicates with the cortex via the mammillothalamic tract - anterior nucleus thalamus - cingulate gyrus route. The amygdala communicates with the hypothalamus via the stria terminalis and the ventral amygdalofugal pathway. Hypothalamic neurons secrete ADH and transport ADH to the neurohypophysis for storage and release as needed. The hypothalamus is the master of both autonomic and endocrine homeostatic and survival systems.

11.006 The hypothalamus has ample connections with the limbic system by virtue of the:
   A. neurohypophysis.
   B. the posterior commissure.
   C. the mammillothalamic tract.
   D. the corpus callosum.
   E. the anterior commissure.

C is the correct answer.
The mammillothalamic tract is the obvious connection, though connections with all hypothalamic feeders (amygdala, hippocampus, etc.) offer ready entry into limbic areas. The other structures are commissural connections between cortical and midbrain regions.

11.007 The amygdala communicates with the hypothalamus by means of:
   A. the mammillothalamic tract.
   B. the corpus callosum.
   C. the stria terminalis.
   D. the septum pellucidum.
   E. the angular gyrus.

C is the correct answer.
The stria terminalis extends from the amygdala to the septal area and to the hypothalamus.

11.008 Select the one INCORRECT statement. The hypothalamus:
   A. merges with the telencephalic paleocortex in the septal-subcallosal area.
   B. controls the autonomic nervous system through its connections with the midbrain reticular system.
   C. is a vital area in the maintenance of homeostasis.
   D. effects the varying visceral accompaniments of varying emotional states.
   E. is concerned with the formation of motor patterns through its connection with the subthalamus.

E is the correct answer.
The hypothalamus is not directly concerned with motor patterns nor does it directly stimulate muscle fibers. The other statements are all accurate.

11.009 One statement below is INCORRECT. The hypothalamus is bounded by, or abuts on:
   A. the cerebral peduncles.
   B. optic tracts.
   C. optic chiasm.
   D. olfactory tracts.
   E. sella turcica.

D is the correct answer.
The olfactory tracts are cranial to the "diamond" formed by the optic chiasm, optic tracts, and the cerebral peduncles. The diamond neatly encloses the basal or ventral aspect of the hypothalamus. The sella turcica containing the pituitary is covered by dura which is obviously pierced by the hypothalamic infundibulum, and is surmounted by the hypothalamic tuber cinereum.

11.010 Select the INCORRECT statement regarding the mammillary body:
   A. Projects to the anterior nucleus of the thalamus.
   B. Is interconnected with the midbrain tegmentum reticular system.
   C. Receives input from the hippocampal formation.
   D. Forms a visible part of the posterior dorsal hypothalamus.
   E. is a diencephalic structure.

D is the correct answer.
Mammillary bodies are visible on the ventral aspect of the diencephalon. The columns of the fornix terminate in large part in the mammillary bodies. A prominent mammillo-tegmental tract connects the two areas.

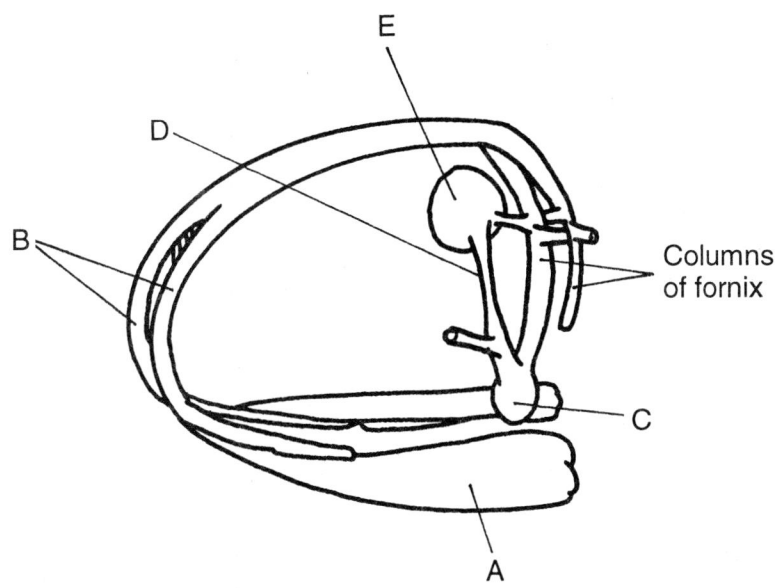

For questions 11 - 16, refer to the diagram above.

11.011 In the diagram above, which is the mammillary body?
C is the correct answer.

11.012 In the diagram above, which projects to cingulate gyrus?
E is the correct answer.

11.013 In the diagram above, which is the anterior nucleus of thalamus?
E is the correct answer.

11.014 In the diagram above, which is the mammillothalamic tract?
D is the correct answer.

11.015 In the diagram above, cells of which structure give origin to fimbria of fornix?
A is the correct answer.

11.016 In the diagram above, which unite to form the body of fornix?
B is the correct answer.

SECTION 12: LIMBIC SYSTEM

12.001 Which of the following structures is/are not part of the limbic system?
   A. Parahippocampal gyrus.
   B. Cingulate gyrus.
   C. Hippocampal formation.
   D. Caudate nucleus.
   E. Septal nuclei.

D is the correct answer.

12.002 The fornix is most closely associated with the:
   A. Amygdala.
   B. Basal ganglia.
   C. Mammillothalamic tract.
   D. Hippocampal formation.
   E. Hypothalamus

D is the correct answer.
The connecting tract between the hippocampus and the septal/hypothalamic areas.

12.003 The uncus is most closely associated with the:
   A. Amygdala.
   B. Basal ganglia.
   C. Mammillothalamic tract.
   D. Hippocampal formation.
   E. Hypothalamus

A is the correct answer.

12.004 The anterior nucleus of thalamus is most closely associated with the:
   A. Amygdala.
   B. Basal ganglia.
   C. Mammillothalamic tract.
   D. Hippocampal formation.
   E. Hypothalamus

C is the correct answer.
Receives the mammillothalamic tract.

12.005 The midbrain reticular nuclei are most closely associated with the:
   A. Amygdala.
   B. Basal ganglia.
   C. Mammillothalamic tract.
   D. Hippocampal formation.
   E. Hypothalamus

E is the correct answer.
These nuclei receive relatively large projections from the hypothalamus (and other limbic structures). Their outflow governs the activity of the autonomic neurons of the cranial nerve nuclei and the spinal cord.

12.006 The archicortex is most closely associated with the:
   A. Amygdala.
   B. Basal ganglia.
   C. Mammillothalamic tract.
   D. Hippocampal formation.
   E. Hypothalamus

D is the correct answer.
Primitive "smell brain" cortex.

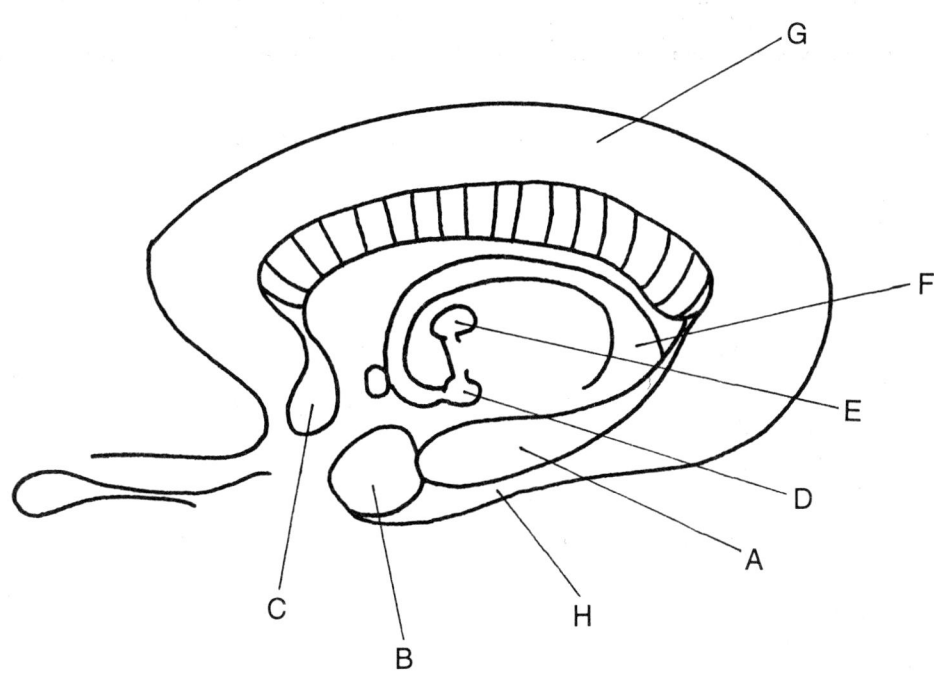

For questions 7 - 13, refer to the diagram above.

12.007 In the diagram above, which structure is the mammillary body?

D is the correct answer.

12.008 In the diagram above, which structure is the hippocampus?

A is the correct answer.

12.009 In the diagram above, which structure is the amygdala?

B is the correct answer.

12.010 In the diagram above, which structure is the septal nucleus?

C is the correct answer.

12.011 In the diagram above, which structure is the cingulate gyrus?

G is the correct answer.

12.012 In the diagram above, which structure is the crus fornix?

F is the correct answer.

12.013 In the diagram above, which structure is the parahippocampal gyrus?

H is the correct answer.

12.014 One of the following statements is INCORRECT. The limbic system:
- A. is considered to be the anatomic substrate underlying behavioral and emotional expression.
- B. plays a role in feelings, feeding, fighting, fleeing, and undertaking mating activity.
- C. plays a major role in the maintenance of a stable body temperature.
- D. expresses itself through the hypothalamus and the autonomic nervous system.
- E. is strongly connected with the hypothalamus by means of the fornix and the stria terminalis.

C is the correct answer.
The hypothalamus alone is capable of maintaining life support systems such as a stable body temperature (through shivering, sweating, skin blood vessel constriction or dilation, etc. - all autonomic system functions directed by hypothalamic temperature sensors). Some consider the hypothalamus and its midbrain reticular autonomic connections part of the limbic system. However, the usual connotation of "limbic" system function is that expressed by statement "A". The involved substrate alluded to in "A" is the anatomic hippocampus-cingulate gyrus-parahippocampal ring, and its subcortical amygdala and septal nuclei.

SECTION 13: MISCELLANEOUS

13.001 Synapses between preganglionic and postganglionic nerve cells:
- A. are associated with dorsal root ganglia only.
- B. are associated with autonomic (visceromotor) ganglia only.
- C. are associated with both dorsal root ganglia and autonomic (visceromotor) ganglia.
- D. are associated with neither dorsal root ganglia nor autonomic (visceromotor) ganglia.

B is the correct answer.

13.002 Which of the following is derived principally from neural crest cells?
- A. dorsal root ganglia only.
- B. autonomic (visceromotor) ganglia only.
- C. both dorsal root ganglia and autonomic (visceromotor) ganglia.
- D. neither dorsal root ganglia nor autonomic (visceromotor) ganglia.

C is the correct answer.

13.003 Pseudounipolar neurons:
- A. are associated with dorsal root ganglia only.
- B. are associated with autonomic (visceromotor) ganglia only.
- C. are associated with both dorsal root ganglia and autonomic (visceromotor) ganglia.
- D. are associated with neither dorsal root ganglia nor autonomic (visceromotor) ganglia.

A is the correct answer.

13.004 Fibrous astrocytes:
- A. are associated with dorsal root ganglia only.
- B. are associated with autonomic (visceromotor) ganglia only.
- C. are associated with both dorsal root ganglia and autonomic (visceromotor) ganglia.
- D. are associated with neither dorsal root ganglia nor autonomic (visceromotor) ganglia.

D is the correct answer.
Fibrous astrocytes are found in the central nervous system, derived from neuroepithelium of the neural tube. Dorsal root and autonomic ganglia are derived from neural crest tissue, which produces peripheral nervous tissue elements.

13.005 A tumor of the pineal gland might:
- A. result in Parkinsonism.
- B. cause hydrocephalus.
- C. result in auditory defects.
- D. cause abnormalities in pupillary light reflexes.
- E. result in gigantism.

B is the correct answer.
Situated as it is, atop the narrow cerebral aqueduct, tumorous enlargement of the pineal may squeeze the aqueduct closed. Continued secretion of the choroid plexi of the lateral and third ventricles then would result in ventricular dilation, obstructive hydrocephalus. The pineal is not related to any of the other statements

13.006 Which of the following is caused by obstruction of the right anterior cerebral artery?
- A. Loss of conscious proprioception in the left leg.
- B. Anesthesia in the right lower face.
- C. Loss of touch in the right leg.
- D. Loss of pain in the upper face, left.
- E. No sensory defect.

A is the correct answer.
The right anterior cerebral artery supplies the medial aspect of the right cerebral hemisphere. This area includes the medial terminations of the primary motor and sensory gyri (paracentral lobule). Represented here are the left foot and leg. Due to crossing of both motor output and sensory input of this area, symptoms will appear on the left side. There will be sensory (and motor) loss in the left foot and leg.

82 Miscellaneous

13.007 Select the discordant pair:
   A. Precentral gyrus - corticobulbar tract.
   B. IX nerve - gag reflex.
   C. Brachium conjunctivum - dentate nucleus.
   D. Visceral sensation - nucleus solitarius.
   E. MLF - medial lemniscus.

E is the correct answer.
Do not confuse the medial longitudinal fasciculus with the medial lemniscus. They are quite separate structures, of entirely different function!

13.008 A stroke involving the genu and posterior portions of the internal capsule on the left will not produce complete paralysis of the right vocal cord because:
   A. The autonomic voice-respiratory center in the reticular substance of the medulla continues to function.
   B. Corticobulbar fibers are in the anterior limb of the internal capsule.
   C. The dorsal motor nucleus of the vagus is controlled by the extrapyramidal system.
   D. Innervation of the nucleus ambiguus is bilateral.
   E. The right vocal cord is innervated by the ventral corticospinal tract of the right side.

D is the correct answer.
Due to bilateral ambiguus innervation, corticobulbar fibers running to it in the internal capsule on one side may be interrupted without deficit in vocal cord function. The dorsal motor nucleus of the vagus and the ventral corticospinal tract are not involved with laryngeal function. Autonomic voice center??!!

13.009 Select the one correct statement:
   A. The nucleus of the oculomotor nerve is located in the central gray matter of the mesencephalon at the level of the inferior colliculus.
   B. In the developing pons of the human fetus, the motor nucleus of the facial nerve is formed from the basal plate.
   C. The corticospinal tract crosses to the opposite side of the central nervous system in the region of the pons.
   D. In the human brain, the pyramids develop from the basal plate of the myelencephalon.
   E. The red nucleus is present in all levels of the midbrain.

B is the correct answer.
The basal plate forms motor neurons. The nucleus of the oculomotor nerve is at the level of the superior colliculus. The corticospinal tract crosses in the lower medulla. The pyramids contain fibers originating in neurons of the cerebral cortex. The red nucleus is only seen in the cranial part of the midbrain, whence it extends somewhat up into the diencephalon.

13.010 Select the one incorrect statement:
   A. Damage to the lateral part of the medulla can cause loss of the sensations of pain and temperature from the opposite side of the body and the same side of the head.
   B. A cerebellar hemisphere influences the musculature of the same side of the body.
   C. The middle cerebellar peduncle consists entirely of pontocerebellar fibers.
   D. Axons of Purkinje cells do not leave the cerebellum.
   E. The diencephalon is separated from the telencephalon laterally by the anterior limb of the internal capsule.

E is the correct answer.
The separation is by the posterior limb of the internal capsule. The lateral medulla houses the crossed spinothalamic tract and the uncrossed descending spinal tract of the trigeminal nerve. This situation produces same side (V) and contralateral side (spinothalamic) losses with lateral medullary injury. One cerebellar hemisphere influences musculature of its side despite its crossing outflow (decussation of the brachium conjunctivum in the midbrain). The subsequent crossing of the descending motor tracts restores cerebellar influence to its side of origin. Purkinje cells terminate in cerebellar roof nuclei. The huge middle cerebellar peduncle does indeed consist of pontocerebellar fibers.

13.011 Select the incorrect statement:
   A. The nucleus dorsalis (Clarke's column) extends through the entire length of the spinal cord.
   B. The blood flow received by the spinal arteries arising from the vertebral arteries is sufficient for only the upper cervical segments of the cord.
   C. The nucleus gracilis is concerned with certain sensations from the homolateral side of the body below the midthoracic level.
   D. The constituent fibers of the medial lemniscus have a somatotopic distribution.
   E. The sensations with which the spinothalamic tract is concerned are said to be "nociceptive" (L. noceo - to hurt).

A is the correct answer.
Clarke's column extends from C8 to L3. Blood supply to the spinal cord must be supplemented at intervals to insure adequate flow (thyrocervical trunk, an occasional radicular arterial branch of an intercostal, the "great radicular artery of Adamkiewicz, etc.). The other three statements are correct.

13.012 Which statement concerning the lateral corticospinal tract is false?
   A. Also known as the pyramidal tract.
   B. Its fibers arise mainly from the frontal lobe.
   C. About 55% of its fibers cross at the lower end of the medulla and enter the opposite side of the cord.
   D. Descends through the entire length of the spinal cord.
   E. A pathway concerned with voluntary, discrete, skilled movements.

C is the correct answer.
Approximately 90% of pyramidal fibers cross at the junction of the medulla and the spinal cord and form the lateral corticospinal tract.

13.013 Select the incorrect statement regarding the results of destructive lesions of lower motor neurons:
   A. Progressive atrophy of the muscles deprived of motor fibers.
   B. Diminished or absent tendon reflexes.
   C. Flaccid paralysis of the affected muscles.
   D. Stroking of the sole causes upturning of the great toe and spreading of the other toes.
   E. No loss of somatic or visceral sensation.

D is the correct answer.
The upturning great toe and spreading of the others is a "primitive" reflex appearing after loss of upper motor neuron control of anterior horn cells. Obviously it requires functioning anterior horn cells - lower motor neurons. (In this case, no motor reaction at all would be seen on stroking the sole.)

13.014 Select the incorrect statement below:
   A. The sulcus limitans divides the alar from the basal plate.
   B. Neural crest tissue gives rise to: all sensory ganglia, autonomic ganglia, adrenal medulla, and neurolemma cells.
   C. With hemisection of the spinal cord, epicritic touch (stereognosis), conscious proprioception, and bone vibratory sense are lost on the side of the lesion. Pain and temperature sensations are lost on the opposite side of the lesion. Motor deficit is found on the same side as the lesion.
   D. The nucleus dorsalis extends from the level of C8 to L3.
   E. Fibers composing the fasciculus cuneatus have cell bodies in dorsal root ganglia of T6 down, on the contralateral side of the cord.

E is the correct answer.
The cuneate fasciculus represents the body from T6 up and on the same side. Hemisection of the cord (Brown-Sequard syndrome) interrupts the crossed descending pyramidal (lateral corticospinal) tract causing motor loss on the side of injury. Ascending uncrossed sensory tracts (gracilis and cuneatus) will be interrupted causing similarly ipsilateral loss (conscious proprioception, stereognosis, etc.). Ascending sensory tracts that have already crossed (spinothalamic tract) will be cut causing loss of pain and temperature sensations from their parent contralateral side. Statements A, B, C, D are correct.

13.015 Select the correct statement:
  A. The GVE cell column of the sacral cord (S2, 3, 4) is large enough to produce a bulge of nuclear material into the lateral funiculus.
  B. The spinal cord terminates usually at the upper border of vertebra L2.
  C. The denticulate ligament is composed of a fold of dura mater extending from the lateral side of the cord (between dorsal and ventral roots) to attach to the parietal dura.
  D. Fibers of the ventral corticospinal tract cross at the lower level of the medulla.
  E. Fibers of second order neurons of the spinothalamic tract cross in the middle level of the medulla.

B is the correct answer.
The sacral GVE column is not large enough to produce a lateral horn as does the thoracolumbar GVE column. The denticulate ligament is formed from the pia mater. Ventral corticospinal fibers do not cross in the medulla with the rest of the corticospinal tract. What crossing they do, is accomplished at the cord level of their synapse with the anterior horn apparatus. Spinothalamic fibers cross in the anterior white commissure of the spinal cord, either at the level of entrance to the cord and/or a segment or two higher, having ascended in the tract of Lissauer.

13.016 The origin of lenticular fasciculus is most closely associated with:
  A. Globus pallidus.
  B. Putamen.
  C. Red nucleus.
  D. Anterior nucleus thalamus (ANT).
  E. Ventral anterior nucleus thalamus (VA).

A is the correct answer.

13.017 The neostriatum is most closely associated with:
  A. Globus pallidus.
  B. Putamen.
  C. Red nucleus.
  D. Anterior nucleus thalamus (ANT).
  E. Ventral anterior nucleus thalamus (VA).

B is the correct answer.
The neostriatum consists of the putamen and the caudate nucleus. The globus pallidus is the paleostriatum.

13.018 The genu of internal capsule is most closely associated with:
  A. Globus pallidus.
  B. Putamen.
  C. Red nucleus.
  D. Anterior nucleus thalamus (ANT).
  E. Ventral anterior nucleus thalamus (VA).

A is the correct answer.
The apex of the lenticular nucleus consists of the tip of the globus pallidus. This presses against the internal capsule causing a bend (genu), with resulting anterior and posterior limbs running to it.

13.019 The projection to cingulate gyrus is most closely associated with:
  A. Globus pallidus.
  B. Putamen.
  C. Red nucleus.
  D. Anterior nucleus thalamus (ANT).
  E. Ventral anterior nucleus thalamus (VA).

D is the correct answer.

13.020 The destination of thalamic fasciculus is most closely associated with:
  A. Globus pallidus.
  B. Putamen.
  C. Red nucleus.
  D. Anterior nucleus thalamus (ANT).
  E. Ventral anterior nucleus thalamus (VA).

E is the correct answer.

13.021 The lenticular fasciculus is most closely associated with:
  A. Globus pallidus.
  B. Caudate nucleus.
  C. Red nucleus.
  D. Substantia nigra.
  E. Ventral posterolateral nucleus thalamus (VPL).

A is the correct answer.

13.022 The neostriatum is most closely associated with:
   A. Globus pallidus.
   B. Caudate nucleus.
   C. Ventral lateral nucleus thalamus (VL).
   D. Substantia nigra.
   E. Ventral posterolateral nucleus thalamus (VPL).

B is the correct answer.

13.023 The dentatothalamic tract (brachium conjunctivum) is most closely associated with:
   A. Globus pallidus.
   B. Caudate nucleus.
   C. Ventral lateral nucleus thalamus (VL).
   D. Substantia nigra.
   E. Ventral posterolateral nucleus thalamus (VPL).

C is the correct answer.

13.024 Paralysis agitans (Parkinson's disease) is most closely associated with:
   A. Globus pallidus.
   B. Caudate nucleus.
   C. Ventral lateral nucleus thalamus (VL).
   D. Substantia nigra.
   E. Ventral posterolateral nucleus thalamus (VPL).

D is the correct answer.

13.025 The spinothalamic tract is most closely associated with:
   A. Globus pallidus.
   B. Caudate nucleus.
   C. Ventral lateral nucleus thalamus (VL).
   D. Substantia nigra.
   E. Ventral posterolateral nucleus thalamus (VPL).

E is the correct answer.

13.026 The brachium of the inferior colliculus is most closely associated with:
   A. Anterior nucleus thalamus (ANT).
   B. Ventral lateral nucleus thalamus .
   C. Medial geniculate body.
   D. Ventral posterolateral and posteromedial nuclei thalamus.
   E. Globus pallidus.

C is the correct answer.

13.027 The subthalamus is most closely associated with:
   A. Anterior nucleus thalamus (ANT).
   B. Ventral lateral nucleus thalamus .
   C. Medial geniculate body.
   D. Ventral posterolateral and posteromedial nuclei thalamus.
   E. Globus pallidus.

E is the correct answer.

13.028 The mammillothalamic tract is most closely associated with:
   A. Anterior nucleus thalamus (ANT).
   B. Ventral lateral nucleus thalamus .
   C. Medial geniculate body.
   D. Ventral posterolateral and posteromedial nuclei thalamus.
   E. Globus pallidus.

A is the correct answer.

13.029 The brachium conjunctivum is most closely associated with:
- A. Anterior nucleus thalamus (ANT).
- B. Ventral lateral nucleus thalamus.
- C. Medial geniculate body.
- D. Ventral posterolateral and posteromedial nuclei thalamus.
- E. Globus pallidus.

B is the correct answer.

13.030 Toothache is most closely associated with:
- A. Anterior nucleus thalamus (ANT).
- B. Ventral lateral nucleus thalamus.
- C. Medial geniculate body.
- D. Ventral posterolateral and posteromedial nuclei thalamus.
- E. Globus pallidus.

D is the correct answer.

13.031 The caudate and putamen are most closely associated with:
- A. Brachium conjunctivum (superior cerebellar peduncle).
- B. Decussation of corticospinal tract.
- C. Neostriatum.
- D. Semicircular canals.
- E. Lenticular fasciculus.

C is the correct answer.

13.032 The dentate nucleus is most closely associated with:
- A. Brachium conjunctivum (superior cerebellar peduncle).
- B. Decussation of corticospinal tract.
- C. Neostriatum.
- D. Semicircular canals.
- E. Lenticular fasciculus.

A is the correct answer.
Supplies the bulk of the fibers of the superior cerebellar peduncle.

13.033 The flocculonodular lobe is most closely associated with:
- A. Brachium conjunctivum (superior cerebellar peduncle).
- B. Decussation of corticospinal tract.
- C. Neostriatum.
- D. Semicircular canals.
- E. Lenticular fasciculus.

D is the correct answer.
The archicerebellum. Balance - a primitive sense.

13.034 The ventral anterior nucleus of the thalamus is most closely associated with:
- A. Brachium conjunctivum (superior cerebellar peduncle).
- B. Decussation of corticospinal tract.
- C. Neostriatum.
- D. Semicircular canals.
- E. Lenticular fasciculus.

E is the correct answer.

13.035 The medullary pyramids are most closely associated with:
- A. Brachium conjunctivum (superior cerebellar peduncle).
- B. Decussation of corticospinal tract.
- C. Neostriatum.
- D. Semicircular canals.
- E. Lenticular fasciculus.

B is the correct answer.

13.036 Spasticity and Babinski reflex are most closely associated with:
- A. Cerebellum.
- B. Anterior horn cells.
- C. Bilaterally innervated.
- D. Contralaterally innervated.
- E. Rubrospinal tracts.

B is the correct answer.
These symptoms appear with release of upper motor neuron control of anterior horn cells.

13.037 Axial movement is most closely associated with:
- B. Anterior horn cells.
- C. Bilaterally innervated.
- D. Contralaterally innervated.
- E. Rubrospinal tracts.

C is the correct answer.
Preserved even with loss of all corticospinals of one side because of the bilateral performance of the ventral corticospinal tracts.

13.038 Distal fine movement is most closely associated with:
- A. Cerebellum.
- B. Anterior horn cells.
- C. Bilaterally innervated.
- D. Contralaterally innervated.
- E. Rubrospinal tracts.

D is the correct answer.
The province of the crossed lateral corticospinal tracts.

13.039 Muscle atrophy and weakness are most closely associated with:
- A. Cerebellum.
- B. Anterior horn cells.
- C. Bilaterally innervated.
- D. Contralaterally innervated.
- E. Rubrospinal tracts.

B is the correct answer.
The results of denervation of muscles subsequent to destruction of anterior horn cells.

13.040 Ataxia is most closely associated with:
- A. Cerebellum.
- B. Anterior horn cells.
- C. Bilaterally innervated.
- D. Contralaterally innervated.
- E. Rubrospinal tracts.

A is the correct answer.
Disorder of the cerebellum produces no sensory loss and no paralysis. Incoordination is the hallmark of cerebellar dysfunction.

13.041 Which of the following projects to the contralateral cerebellum?
- A. Nucleus gracilis.
- B. Salivatory nucleus.
- C. Inferior olivary nucleus.
- D. Substantia nigra.
- E. Solitary tract and nucleus.

C is the correct answer.
Exact function of this large cerebellar input is uncertain.

13.042 Which of the following participates in the formation of the medial lemniscus?
- A. Nucleus gracilis.
- B. Salivatory nucleus.
- C. Inferior olivary nucleus.
- D. Substantia nigra.
- E. Solitary tract and nucleus.

A is the correct answer.
Helps form the medial lemniscus through its internal arcuate fibers.

13.043 Which of the following projects to the VPL/VPM of the thalamus?
- A. Nucleus gracilis.
- B. Salivatory nucleus.
- C. Inferior olivary nucleus.
- D. Substantia nigra.
- E. Solitary tract and nucleus.

A is the correct answer.
E is also correct. Gustatory nucleus of the nucleus solitarius does project to the VPM and consciousness.

88   Miscellaneous

13,044   Which of the following projects dopaminergic fibers to the striatum?
   A. Nucleus gracilis.
   B. Salivatory nucleus.
   C. Inferior olivary nucleus.
   D. Substantia nigra.
   E. Solitary tract and nucleus.

D is the correct answer.

13.045   Which of the following projects to the pterygopalatine ganglion?
   A. Nucleus gracilis.
   B. Salivatory nucleus.
   C. Inferior olivary nucleus.
   D. Substantia nigra.
   E. Solitary tract and nucleus.

B is the correct answer.
The salivatory nucleus is the origin of all parasympathetic GVE's going to head structures, save for these from the Edinger-Westphal nucleus going to the eyes. (The dorsal motor nucleus of the vagus does not project to structures in the head.)

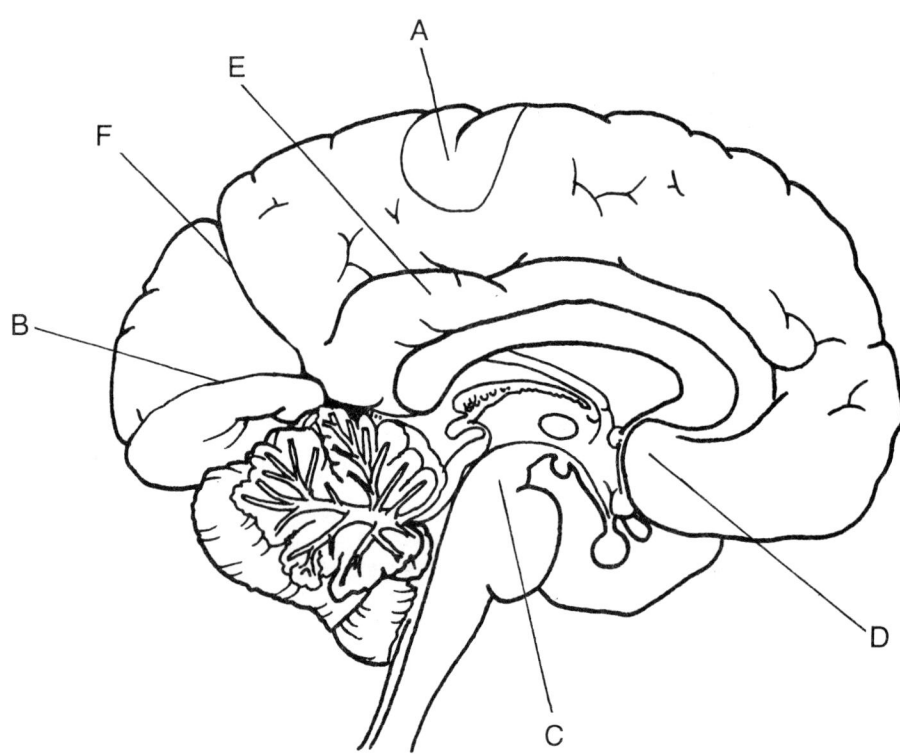

For questions 46 - 50, refer to the diagram above.

13.046   In the diagram above, the terminal portion of the primary sensory and motor strips is labeled _____.

A is the correct answer.

13.047   In the diagram above, _____ is the major cortex of the limbic lobe.

E is the correct answer.

13.048   In the diagram above, _____ is the fissure about which the visual cortex is arranged.

B is the correct answer.

13.049   In the diagram above, _____ is the part of the reticular system intimately connected with the hypothalamus.

C is the correct answer.

13.050   In the diagram above, _____ is the septal area.

D is the correct answer.

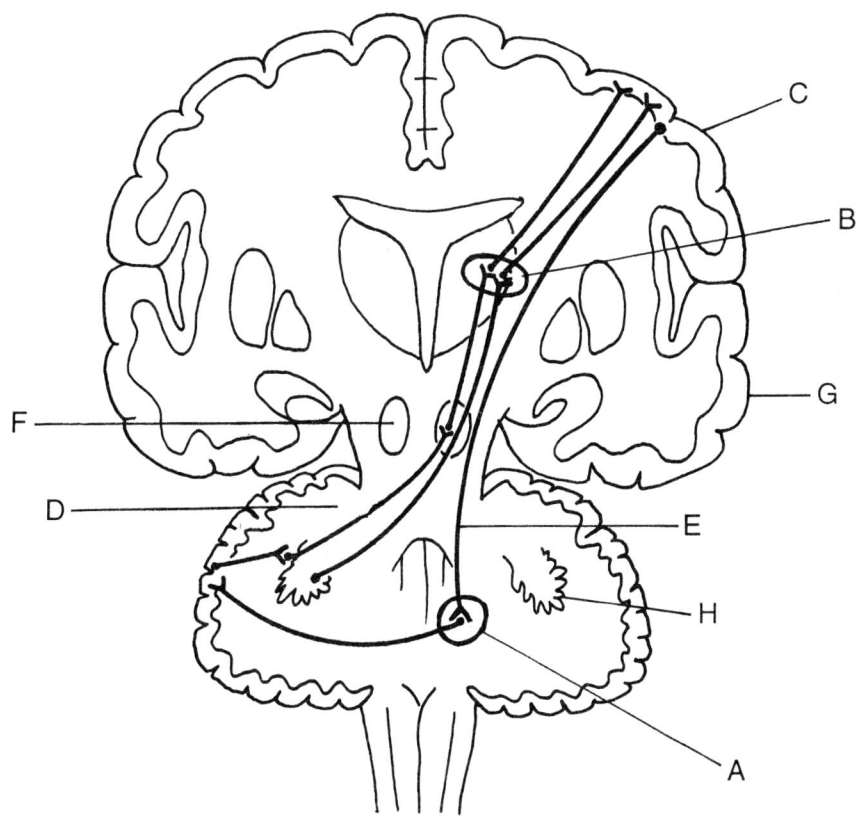

For questions 51 - 55, refer to the diagram above.

13.051 In the diagram above, ____ is corticopontine fibers.   E is the correct answer.

13.052 In the diagram above, ____ is the premotor cortex.   C is the correct answer.

13.053 In the diagram above, ____ is the VL nucleus.   B is the correct answer.

13.054 In the diagram above, ____ is the brachium conjunctivum.   D is the correct answer.

13.055 In the diagram above, ____ is the pontine nucleus.   A is the correct answer.

90   Miscellaneous

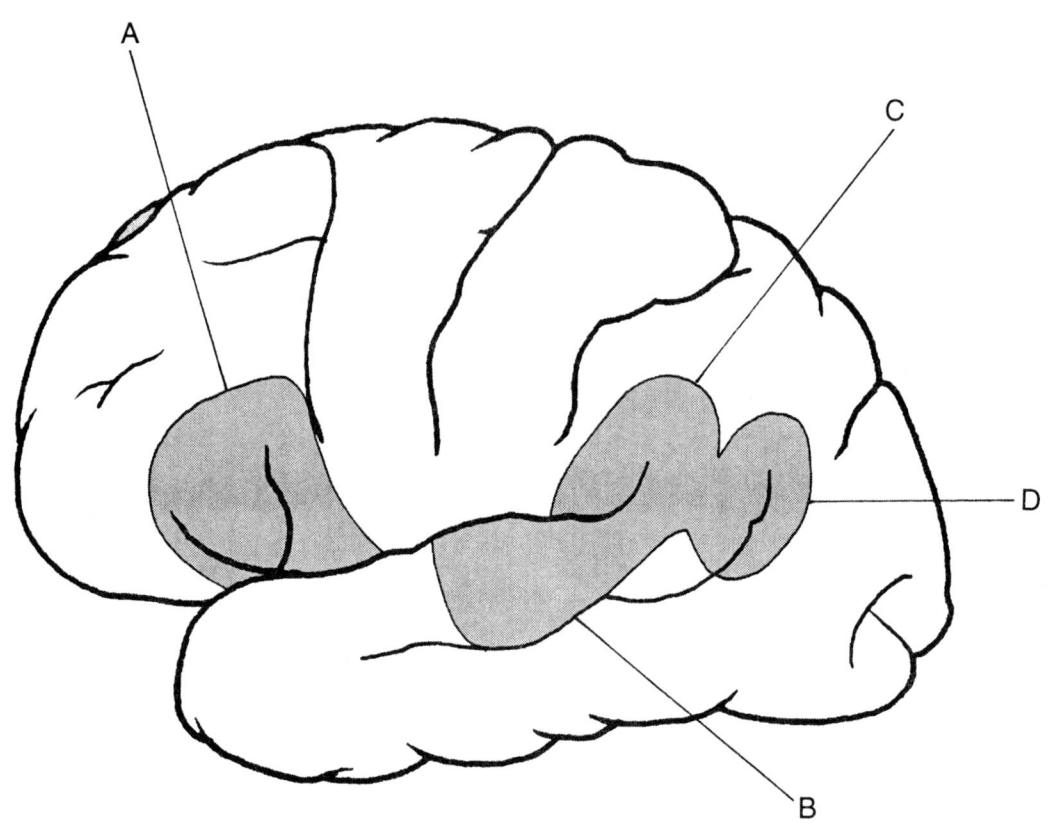

For questions 56 - 60, refer to the diagram above.

13.056 In the diagram above, _____ is Broca's motor speech area.

A is the correct answer.

13.057 In the diagram above, _____ is the supramarginal gyrus (associated with understanding of the spoken word).

C is the correct answer.

13.058 In the diagram above, _____ is the angular gyrus (associated with understanding of the printed word).

D is the correct answer.

13.059 In the diagram above, _____ is the area surrounding the transverse gyrus of Heschl (auditory association cortex.

B is the correct answer.

13.060 In the diagram above, the lateral geniculate body would have most influence on which lettered area?

D is the correct answer.

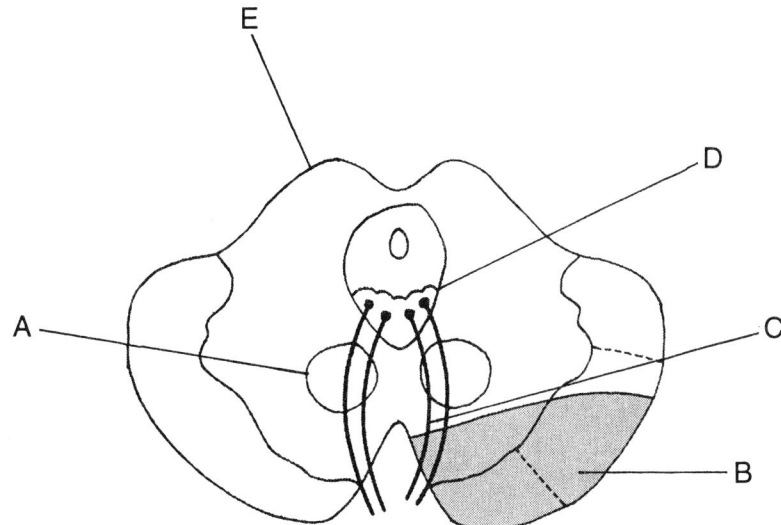

For questions 61 - 67, refer to the diagram above.

13.061 In the diagram above, _____ is the red nucleus.

A is the correct answer.

13.062 In the diagram above, _____ is the cerebral peduncle.

B is the correct answer.
Corticospinal fibers are found in the central portion of this structure.

13.063 In the diagram above, _____ is the oculomotor nerve.

C is the correct answer.
Emerges characteristically from the interpeduncular fossa as shown.

13.064 In the diagram above, _____ is the oculomotor nerve nucleus.

D is the correct answer.
The Edinger-Westphal nucleus is part of this complex.

13.065 In the diagram above, _____ is the superior colliculus.

E is the correct answer.
Always found at the same level as the oculomotor nerve nucleus.

13.066 Which one of the following statements is INCORRECT, given the hatched lesion in the diagram above?
  A. Contralateral hemiparesis due to interruption of corticospinal fibers.
  B. Ipsilateral paralysis of most ocular muscles.
  C. Ipsilateral outward turning of the eye.
  D. Constriction of pupil due to interruption of parasympathetic fibers.
  E. Ptosis (drooping of the upper eyelid) would also be observed.

D is the correct answer.
With loss of Edinger-Westphal parasympathetics, the sympathetics to the eye are unopposed and will cause dilation of the ipsilateral pupil. The superior oblique and the lateral rectus muscles have their own nerves (trochlear and abducens). All the other extraocular muscles will be paralyzed, including the levator palpebrae which is also innervated by the oculomotor. The injury to the oculomotor is peripheral, there will be no more crossing. The symptoms will be ipsilateral. The affected eye will be turned outward due to unopposed pull of the lateral rectus muscle.

13.067 What artery might be damaged and cause the hatched lesion in the diagram above?
  A. Anterior cerebral artery
  B. Middle cerebral artery
  C. Posterior cerebral artery
  D. Anterior communicating artery
  E. Posterior communicating artery

C is the correct answer.
The posterior cerebral artery. The oculomotor nerve is directly subjacent to the posterior cerebral as the latter winds around the cerebral peduncle. This is the not uncommon Weber's syndrome.

13.068 A 50-year old diabetic with hypertension, who smokes heavily (three dispositions to atherosclerosis) is admitted to the hospital with right-sided weakness involving both the arm and leg. The patient's deep tendon reflexes are hyperactive and a right Babinski sign is elicited. Cranial nerve examination reveals a striking weakness of the left side of the tongue causing slurred speech and deviation to the left when protruded. You quickly surmise that the responsible stroke is localized in the:
- A. Left cerebral artery territory.
- B. Left internal capsule.
- C. Right cervical spinal cord.
- D. Left paramedian medulla with destruction of the pyramid and proximal XII nerve.
- E. Right cerebral peduncle.

D is the correct answer.
In this location, the pyramids have not yet decussated - their interruption here will therefore produce contralateral paralysis. The XII nerve damage is peripheral - no more crossing. Symptoms of left tongue palsy include slurred speech and protrusion to the left - the side of injury (paralysis of the left genioglossus). No other location will produce this crossed ("alternating") palsy.

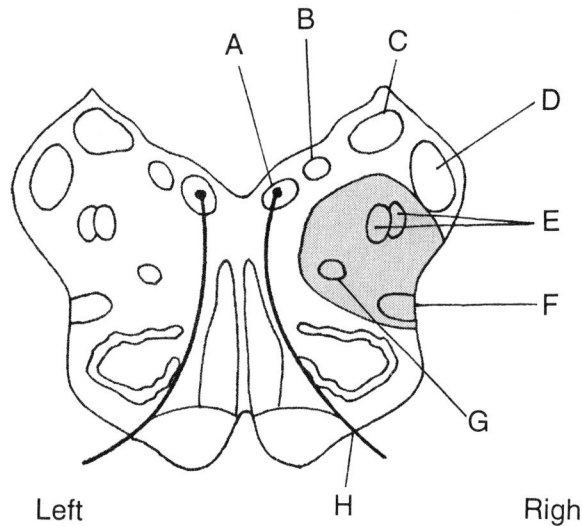

For questions 69 - 76, refer to the diagram above.

13.069 In the diagram above, _____ is the hypoglossal nucleus.

A is the correct answer.

13.070 In the diagram above, _____ is the dorsal motor nucleus of vagus.

B is the correct answer.

13.071 In the diagram above, _____ is the vestibular nucleus.

C is the correct answer.

13.072 In the diagram above, _____ is the inferior cerebellar peduncle.

D is the correct answer.

13.073 In the diagram above, _____ is the spinal tract and nucleus of trigeminal.

E is the correct answer.

13.074 In the diagram above, _____ is the spinothalamic tract.

F is the correct answer.

13.075 In the diagram above, _____ is the nucleus ambiguus.

G is the correct answer.

13.076 In the diagram above, _____ is the hypoglossal nerve.

H is the correct answer.

13.077 The hatched area of destruction in the diagram is likely produced by:
- A. posterior inferior cerebellar artery.
- B. posterior cerebral artery.
- C. anterior spinal artery.
- D. anterior inferior cerebellar artery.
- E. basilar artery.

A is the correct answer.
This is the classical posterior inferior cerebellar artery (PICA) lesion. The posterior cerebral artery is much more cranial as it winds around the cerebral peduncle of the midbrain to reach the posterior cerebrum. The anterior spinal artery is a branch of the vertebral as is the PICA, but runs caudally along the spinal cord. The anterior inferior cerebellar artery (AICA) supplies the anterior inferior aspect of the cerebellum. The basilar artery is not formed until the pons is reached.

13.078 Symptoms produced by the hatched lesion might include:
- A. ipsilateral paralysis of the tongue.
- B. contralateral loss of pain and temperature sensation in the face.
- C. dysphagia and voice change.
- D. Horner's syndrome left.
- E. Ipsilateral bodily loss of pain and temperature sensation.

C is the correct answer.
The nucleus ambiguus supplies innervation of the branchiogenic muscles of the larynx and pharynx. Difficulty in swallowing and voice change are to be expected with its destruction. Horner's syndrome on the side of the lesion (right) would occur as sympathetic fibers in the medullary tegmentum bound for T1 lateral horn cells are interrupted - ipsilateral Horner's rather than contralateral. Facial pain and temperature loss would be ipsilateral rather than contralateral as trigeminal spinal tract fibers are not yet crossed at the locus of the lesion pictured. Conversely, bodily sensation of pain and temperature will be lost on the contralateral side as the spinothalamic fibers here have already crossed. The tongue nucleus and nerve are shown to be outside the area of the lesion.

13.079 A 25 year-old woman comes to you complaining of sudden onset and persistence of dizziness. She notes that the world seems to be spinning to the left. On examination you note the presence of horizontal nystagmus. There are no other abnormalities. What structure is involved?
- A. MLF.
- B. Edinger-Westphal nucleus.
- C. Oculomotor nerve.
- D. Vestibular apparatus.
- E. Brachium pontis.

D is the correct answer.
The vestibular apparatus is responsible for eye movement in response to head-turning stimuli. In order to move horizontally in conjugate fashion as in this instance, an intact MLF, abducens and oculomotor nerves and nuclei are required. The brachium pontis and Edinger-Westphal nucleus are not concerned.

13.080 A 25 year-old woman comes to you complaining of sudden onset and persistence of dizziness. She notes that the world seems to be spinning to the left. On examination you note the presence of horizontal nystagmus with the fast checking component directed to the left. There are no other abnormalities. On which side is the structure involved?
- A. Right.
- B. Left.
- C. Both.
- D. Neither.

A is the correct answer.
This patient has a problem in her vestibular apparatus. It is emitting false signals indicating a turning to the right (world turning left). This could be due to an irritant disorder in the right horizontal canal causing increased firing and a slow horizontal eye movement to the left ("following the world"); or a disorder in the right vestibular apparatus knocking out its ability to fire. Now the left side is unopposed and acts as if stimulated by a turn to the left. It therefore drives the eyes to the right. In the first case, the fast checking saccadic movement would be to the right. In the second case, it would be to the left. In the patient described, the latter situation exists.

94   Miscellaneous

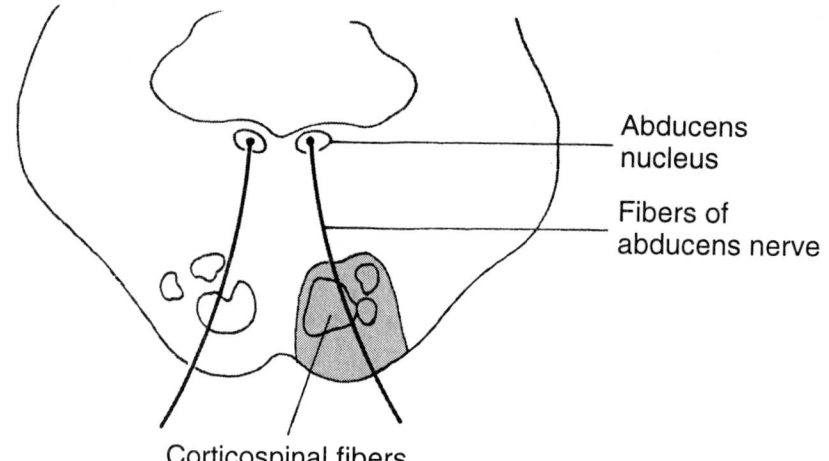

For question 81, refer to the diagram above.

13.081 Select the INCORRECT statement regarding the above diagram.
   A. This lesion would produce symptoms on both sides of the body.
   B. Interruption of corticospinal fibers causes contralateral hemiparesis.
   C. Involvement of abducens nerve would cause paralysis of the lateral rectus muscle on the ipsilateral side, and a medial strabismus (squint).
   D. The corticopontine involvement would not be clinically detectable due to extensive paralysis caused by the loss of the corticospinal tract.
   E. This lesion might be caused by occlusion of a pontine branch of the basilar artery.

D is the correct answer.
A nonsensical statement! The combination of hemiparesis on one side and lateral rectus paralysis on the other accurately pinpoints this anatomic area as the site of trouble. Corticospinal tract in this location is bound for the contralateral musculature, but has not yet crossed (decussation of the pyramids). Abducens nerve is bound for the ipsilateral orbital lateral rectus muscle. Its fibers will cross no more. Loss of the lateral rectus function results in an unopposed pull of the medial rectus which turns the eye medially. The basilar artery passes in the groove shown in the diagram. Its pontine branches supply the lesion area.

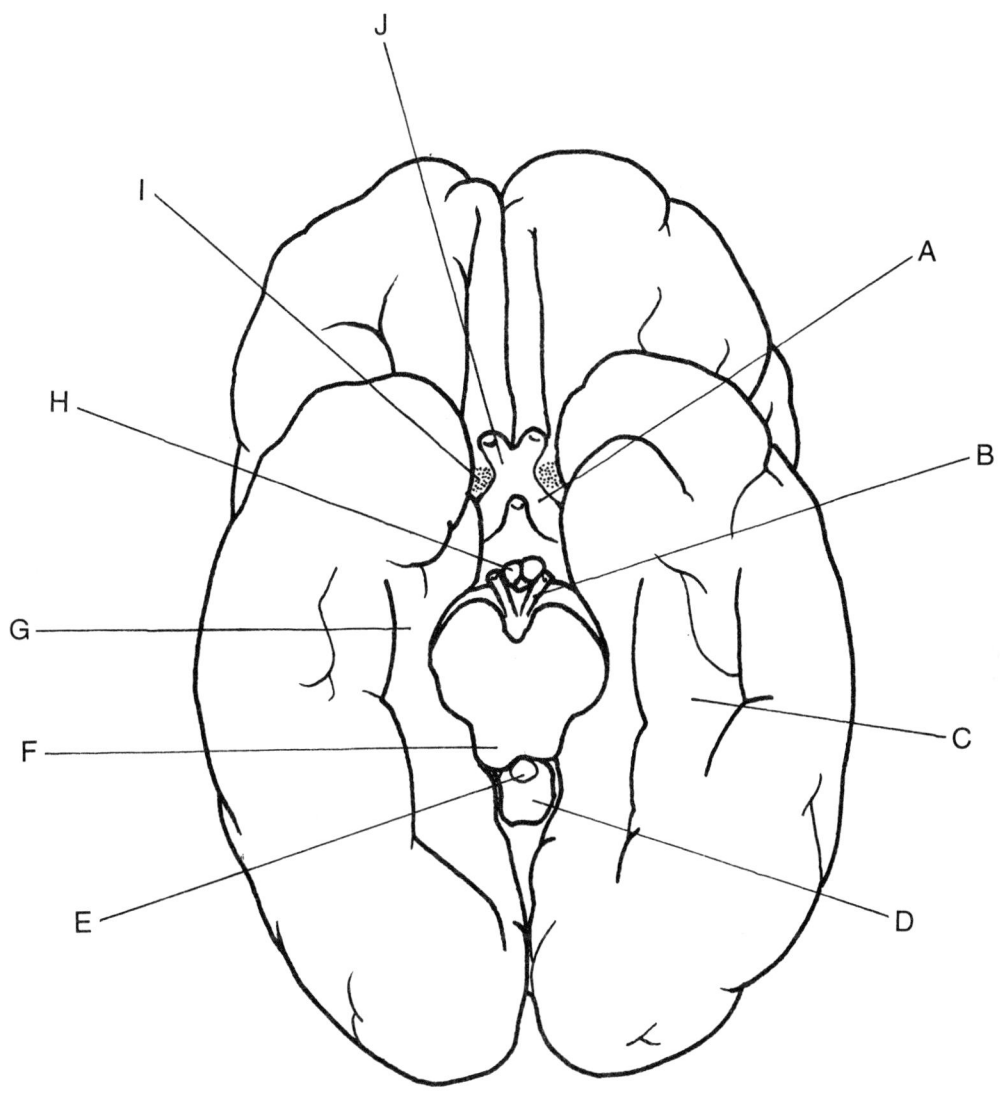

For questions 83 - 86, refer to the diagram above.

13.082 In the diagram above, _____ contains fibers innervating the sphincter muscle of the pupil.

B is the correct answer.
The oculomotor nerve containing parasympathetic preganglionic fibers from its Edinger-Westphal nucleus is found in this position - issuing from the depths of the interpeduncular fossa. After synapse in the ciliary ganglion, postganglionics travel to and innervate the sphincter pupillae.

13.083 In the diagram above, _____ is a diencephalic structure of the limbic system.

H is the correct answer.
The mammillary bodies are a part of the limbic loop: hippocampal fornix - mammillary bodies - anterior nucleus of thalamus - cingulate gyrus.

96   Miscellaneous

13.084 In the diagram above, ____ is an olfactory association area.

G is the correct answer.
The parahippocampal gyrus associates odors with past "odor experiences" enabling full perception and evaluation of the olfactory stimuli.

13.085 In the diagram above, ____ overlies the septal area.

I is the correct answer.
The anterior perforated substance is the "surface" of the septal area.

13.086 In the diagram above, ____ is part of the midbrain.

F is the correct answer.
The superior colliculus - at the level of emergence of the oculomotor nerve.

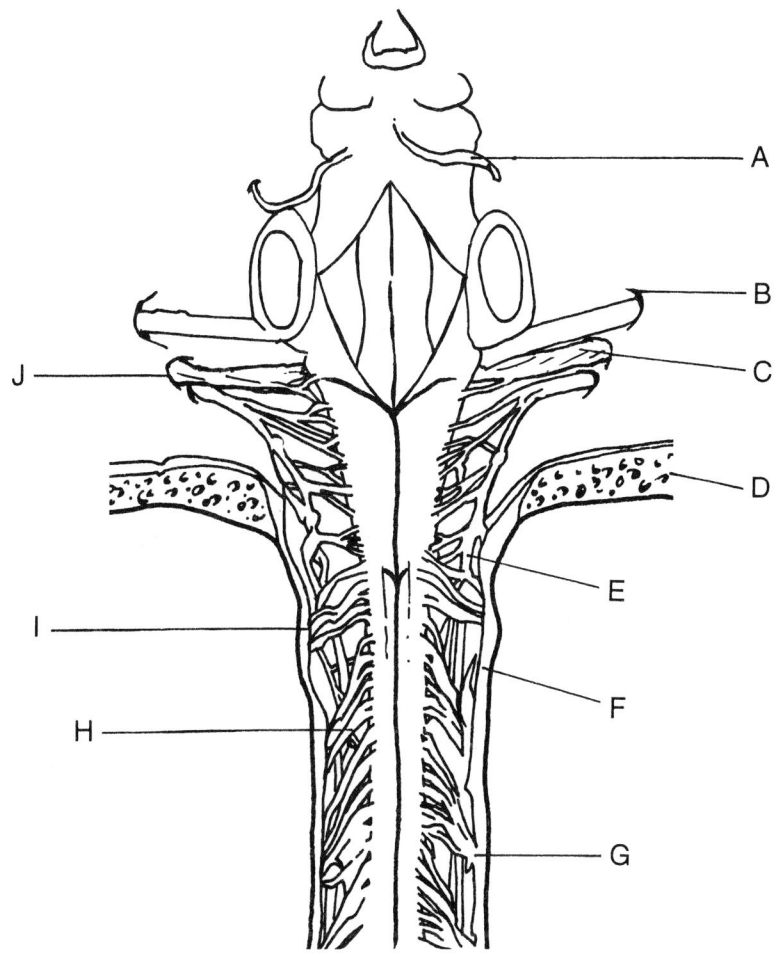

For questions 87 - 91, refer to the diagram above.

13.087 In the diagram above, _____ is the dura mater.

F is the correct answer.
Dura is loose around the spinal cord, but as seen in the figure, is fused with the periosteum of the cranial vault.

13.088 In the diagram above, _____ is the internal auditory meatus.

B is the correct answer.

13.089 In the diagram above, _____ is the spinal portion of spinal accessory nerve.

E is the correct answer.
Spinal accessory nerve receives input from as low as C5.

13.090 In the diagram above, _____ is the trochlear nerve.

A is the correct answer.
The only cranial nerve to issue from the dorsum of the brainstem.

13.091 In the diagram above, _____ is the sensory rootlets of C2.

I is the correct answer.
This is the dorsal aspect of the brainstem and cord. The rootlets visible are dorsal sensory rootlets. C1 has no sensory component; thus this is C2.

98 Miscellaneous

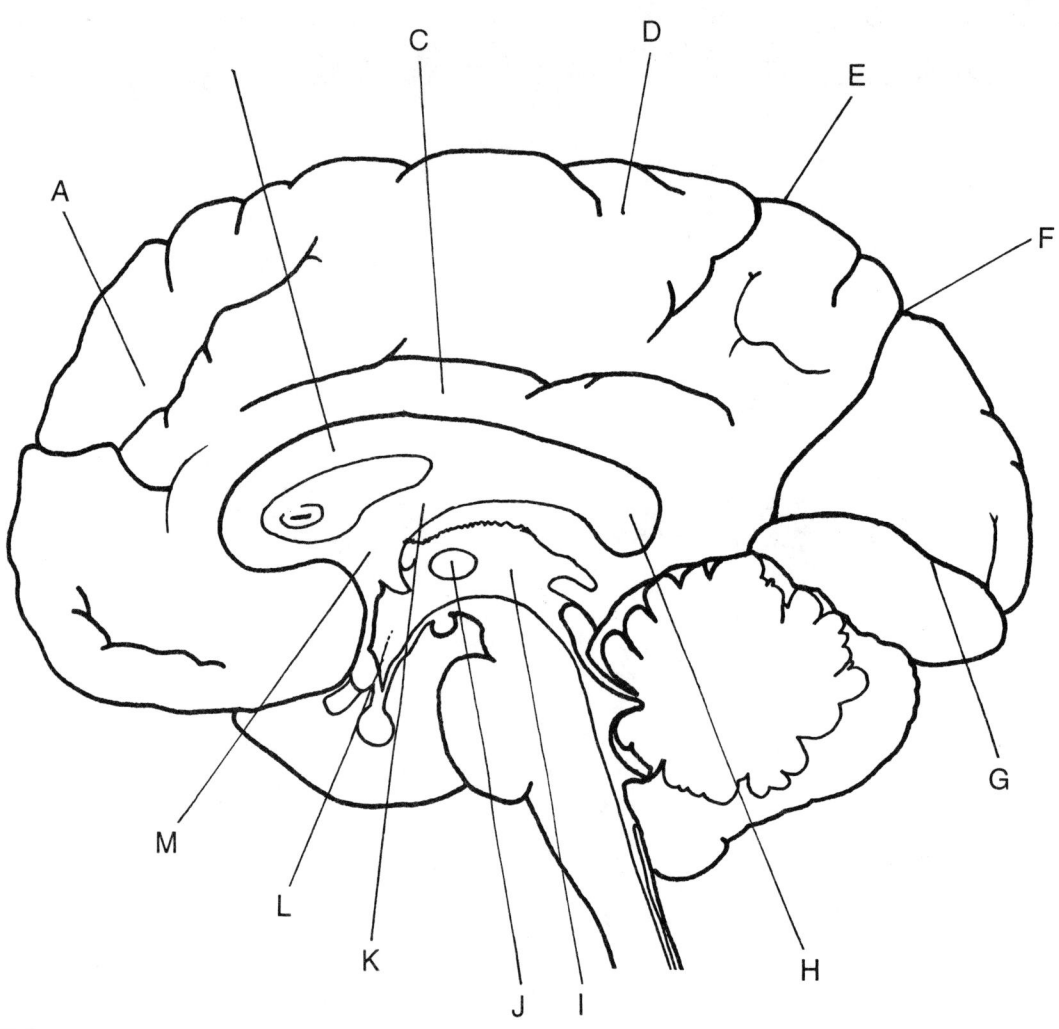

For questions 92 - 96, refer to the diagram above.

13.092 In the diagram above, _____ is the cortex of the limbic lobe.

C is the correct answer.
Cortex of the limbic lobe is the cingulate gyrus.

13.093 In the diagram above, _____ is an area housing autonomic system nuclei.

L is the correct answer.
The hypothalamus.

13.094 In the diagram above, _____ is the body of the corpus callosum.

B is the correct answer.

13.095 In the diagram above, _____ is the sulcus separating the parietal and occipital lobes.

F is the correct answer.
The parieto-occipital sulcus.

13.096 In the diagram above, _____ is the communicating tract between hippocampus and hypothalamus.

K is the correct answer.
The fornix.

SECTION 14: SPINAL CORD

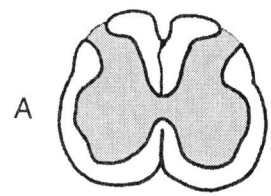
A

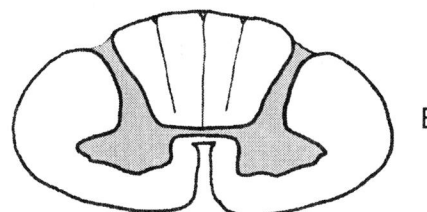
B

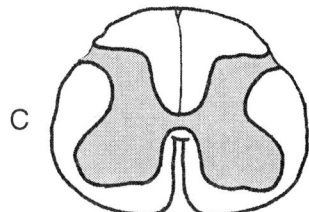
C

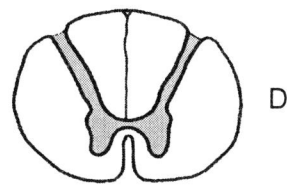
D

For questions 1 - 4, refer to the diagram above.

14.001 In the diagram above, _____ is L3 level.   C is the correct answer.

14.002 In the diagram above, _____ is T10 level.   D is the correct answer.

14.003 In the diagram above, _____ is C7 level.   B is the correct answer.

14.004 In the diagram above, _____ is S3 level.   A is the correct answer.

14.005 Nucleus ventralis posteromedialis is most closely   C is the correct answer.
associated with:
    A. Nucleus cuneatus.
    B. Hair follicle plexus.
    C. Second order trigeminal fibers.
    D. Zone of Lissauer.

14.006 Light touch is most closely associated with: B is the correct answer.
    A. Nucleus cuneatus.
    B. Hair follicle plexus.
    C. Second order trigeminal fibers.
    D. Zone of Lissauer.

14.007 The dorsal portion of medial lemniscus in medulla is   A is the correct answer.
most closely associated with: Upper body here, from nucleus cuneatus.
    A. Nucleus cuneatus.
    B. Hair follicle plexus.
    C. Second order trigeminal fibers.
    D. Zone of Lissauer.

14.008 Pain and temperature are most closely associated   D is the correct answer.
with: Nucleus cuneatus, conscious proprioception.
    A. Nucleus cuneatus.   Nucleus dorsalis, unconscious proprioception.
    B. Hair follicle plexus.   Hair follicle plexus, light touch.
    C. Nucleus dorsalis.
    D. Zone of Lissauer.

14.009 Which of the following is located in the basal plate of the spinal cord?
   A. Postganglionic sympathetic neuron.
   B. Postganglionic parasympathetic neuron.
   C. General somatic afferent (GSA) neuron.
   D. General somatic efferent (GSE) neuron.
   E. None of the above.

D is the correct answer.

14.010 Which of the following is located in the dorsal root ganglion of the cord?
   A. Postganglionic sympathetic neuron.
   B. Preganglionic parasympathetic neuron.
   C. General somatic afferent (GSA) neuron.
   D. General somatic efferent (GSE) neuron.
   E. None of the above.

C is the correct answer.

14.011 Which of the following is most closely associated with the nucleus dorsalis?
   A. Postganglionic sympathetic neuron.
   B. Preganglionic parasympathetic neuron.
   C. General somatic afferent (GSA) neuron.
   D. General somatic efferent (GSE) neuron.
   E. None of the above.

E is the correct answer.
Nucleus dorsalis receives axons from muscle spindles and tendon organs (GVA).

14.012 Which of the following is most closely associated with the sympathetic trunk?
   A. Postganglionic sympathetic neuron.
   B. Preganglionic parasympathetic neuron.
   C. General somatic afferent (GSA) neuron.
   D. General somatic efferent (GSE) neuron.
   E. None of the above.

A is the correct answer.

14.013 Which of the following is most closely associated with the celiac ganglion?
   A. Postganglionic sympathetic neuron.
   B. Preganglionic parasympathetic neuron.
   C. General somatic afferent (GSA) neuron.
   D. General somatic efferent (GSE) neuron.
   E. None of the above.

A is the correct answer.

14.014 Which of the following is most closely associated with the lateral horn.?
   A. Postganglionic sympathetic neuron.
   B. Preganglionic parasympathetic neuron.
   C. General somatic afferent (GSA) neuron.
   D. General somatic efferent (GSE) neuron.
   E. None of the above.

E is the correct answer.
(See 14.015

14.015 General visceral efferent would most closely apply to:
   A. Anterior horn cells.
   B. Fasciculus cuneatus.
   C. Substantia gelatinosa.
   D. Lateral horn.
   E. Nucleus dorsalis (Clarke).

D is the correct answer.

14.016 Spinocerebellum would most closely apply to:
   A. Anterior horn cells.
   B. Fasciculus cuneatus.
   C. Substantia gelatinosa.
   D. Lateral horn.
   E. Nucleus dorsalis (Clarke).

E is the correct answer.

14.017 Gamma loop would most closely apply to:
A. Anterior horn cells.
B. Fasciculus cuneatus.
C. Substantia gelatinosa.
D. Lateral horn.
E. Nucleus dorsalis (Clarke).

A is the correct answer.
(Also communicates with the cerebellum presumably through the nucleus dorsalis as one possible route.)

14.018 Nucleus spinal tract trigeminal nerve would most closely apply to:
A. Anterior horn cells.
B. Fasciculus cuneatus.
C. Substantia gelatinosa.
D. Lateral horn.
E. Nucleus dorsalis (Clarke).

C is the correct answer.
(The two are similar cell columns handling the same modalities and projecting to related thalamic nuclear areas.)

14.019 Stereognosis would most closely apply to:
A. Anterior horn cells.
B. Fasciculus cuneatus.
C. Substantia gelatinosa.
D. Lateral horn.
E. Nucleus dorsalis (Clarke).

B is the correct answer.

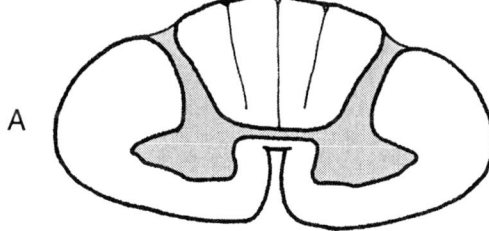
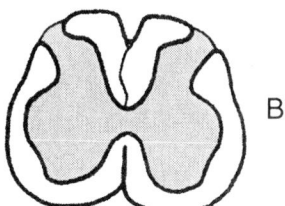

For questions 20 - 23, refer to the diagrams of transverse cord sections above.

14.020 Which of the following contains fibers of dorsal spinocerebellar tract?
A. Lower cervical spinal cord (A) only.
B. Sacral spinal cord (B) only.
C. Lower cervical spinal cord (A) and Sacral spinal cord (B).
D. Neither lower cervical spinal cord (A) nor Sacral spinal cord (B)

A is the correct answer.
Dorsal spinocerebellar tract begins at L3 as the neurons of its fibers originate in the column of Clarke. Clarke's column extends down only to the L3 level.

14.021 Contains preganglionic sympathetic neurons.
A. Lower cervical spinal cord (A) only.
B. Sacral spinal cord (B) only.
C. Lower cervical spinal cord (A) and Sacral spinal cord (B).
D. Neither lower cervical spinal cord (A) nor Sacral spinal cord (B)

D is the correct answer.
The GVE sympathetic neurons are present only from T1 to L2.

14.022 The more rostral of the two sections.
A. Lower cervical spinal cord (A) only.
B. Sacral spinal cord (B) only.
C. Lower cervical spinal cord (A) and Sacral spinal cord (B).
D. Neither lower cervical spinal cord (A) nor Sacral spinal cord (B)

A is the correct answer.

14.023 Contains ascending branches of dorsal root ganglion cells.
A. Lower cervical spinal cord (A) only.
B. Sacral spinal cord (B) only.
C. Lower cervical spinal cord (A) and Sacral spinal cord (B).
D. Neither lower cervical spinal cord (A) nor Sacral spinal cord (B)

C is the correct answer.
Wherever there are receptors for pain-temperature-touch, from the perianal area to the occipital scalp, there is a dorsolateral tract of Lissauer.

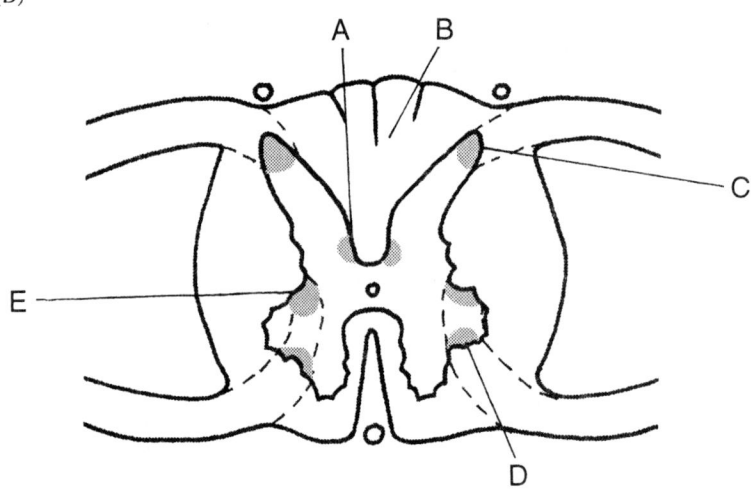

For questions 24 - 28, refer to the diagram above.

14.024 In the diagram above, ____ is a sensory nucleus communicating with the cerebellum.

A is the correct answer.
The dorsal thoracic nucleus (Clarke); the cells of origin of the dorsal spinocerebellar tract.

14.025 In the diagram above, ____ is a sensory nucleus projecting to the VPL of thalamus.

C is the correct answer.
The substantia gelatinosa, whose neurons provide the spinothalamic tract.

14.026 In the diagram above, ____ is a nucleus receiving sensory input from muscle spindles and Golgi tendon organs.

A is the correct answer.
A large responsibility of the column of Clarke and its dorsal spinocerebellar tract is to communicate to the cerebellum continuously the state of muscle tone.

14.027 In the diagram above, ____ is a motor nucleus connected with the reticular system of the midbrain.

E is the correct answer.
GVE neurons of the lateral horn are supervised by reticular system tracts stemming from the hypothalamus via midbrain tegmental reticular centers.

14.028 In the diagram above, ____ contains cells of origin of gamma motor fibers supplying the intrafusal muscle fibers of the muscle spindles.

D is the correct answer.
Gamma motor neurons as well as the more "popular" alpha motor neurons are bona fide inhabitants of the anterior horn.

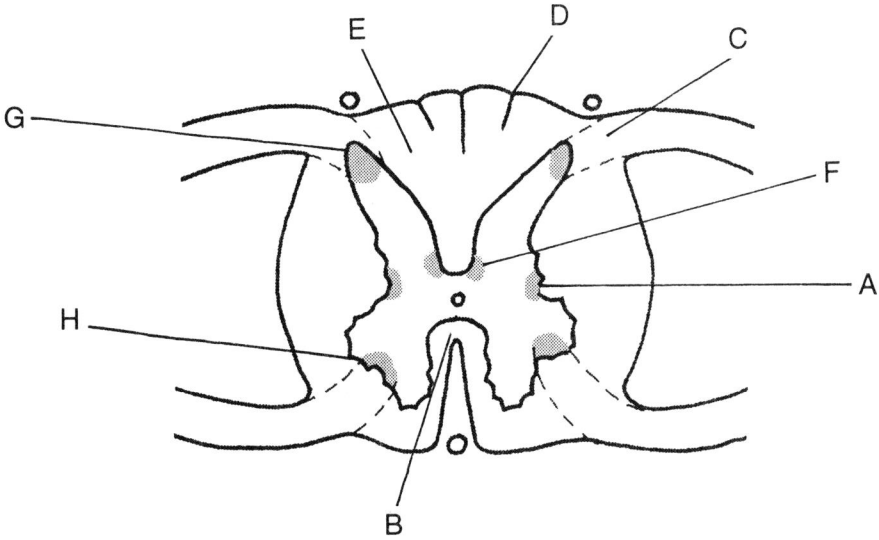

For questions 29 - 35, refer to the diagram above.

14.029 In the diagram above, ____ is a large lateral GVE column.

A is the correct answer.
The GVE cell column produces the visible lateral horn in the thoracic cord. (The parasympathetic GVE cell column in the S2-3-4 cord segments does not contain sufficient neurons to produce a visible S2-3-4 cord swelling.)

14.030 In the diagram above, ____ is the dorsal intermediate septum.

D is the correct answer.
The septum demarcating the gracilis and cuneate fasciculi lies in an intermediate position between the dorsal median and dorsolateral sulci.

14.031 In the diagram above, ____ is the dorsolateral fasciculus (of Lissauer).

C is the correct answer.
Incoming sensory fibers may run up or down a spinal segment or two before plunging into the central cord.

14.032 In the diagram above, ____ contains fibers recording pain and temperature stimuli.

B is the correct answer.
Fibers of the spinothalamic tract decussate here.

14.033 In the diagram above, fibers in fasciculus ____ have their cell bodies located in dorsal root ganglia above the level of T6.

E is the correct answer.
Above T6, incoming conscious proprioceptive fibers form the cuneate fasciculus.

14.034 In the diagram above, ____ contains cell bodies of origin of the dorsal spinocerebellar tract.

F is the correct answer.
The GVA cell column of Clarke, the dorsal thoracic nucleus.

14.035 In the diagram above, ____ is the substantia gelatinosa.

G is the correct answer.
Second order neurons for incoming pain and temperature fibers form this nucleus. Their fibers form the spinothalamic tract which progresses to the third order neurons in the thalamus. (Cf, the spinal tract and nucleus of the trigeminal).

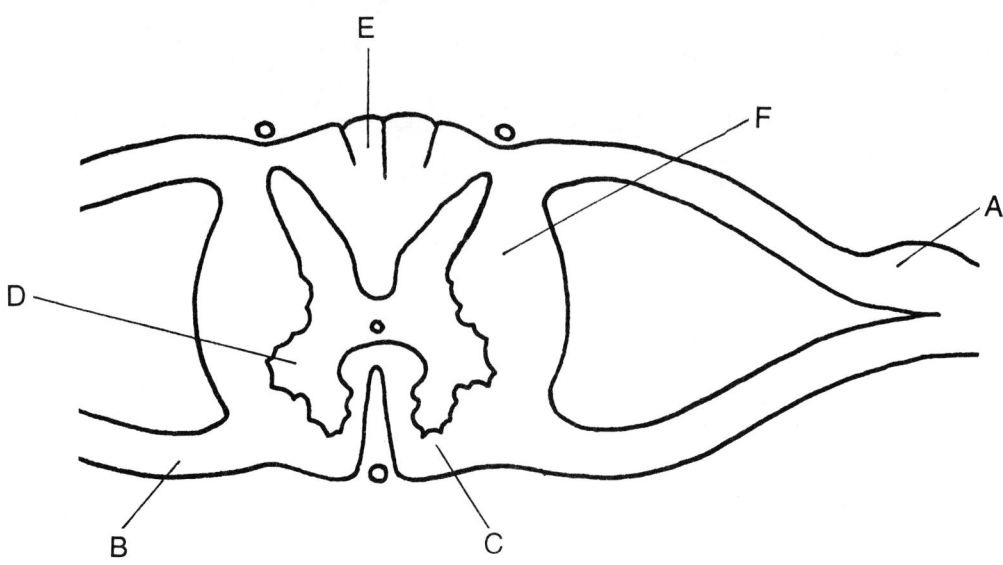

For questions 36 - 40, refer to the diagram above.

14.036 In the diagram above, ____ contains motor fibers running to deep back muscles.

B is the correct answer.
The dorsal primary ramus branches off from the formed nerve, i.e., lateral to the junction of anterior and posterior roots. Its motor fibers are contained in the anterior root.

14.037 In the diagram above, ____ is the anterior funiculus.

C is the correct answer.

14.038 In the diagram above, ____ contains stereognostic information from receptors in the lower extremities.

E is the correct answer.
This is upper thoracic cord with both gracilis and cuneate fasciculi. The medial element of the posterior columns - gracilis - carries stereognostic information from the lower extremities. The dorsal root ganglion contains cell bodies of stereognostic receptors, but at this cord level, its central processes will enter the lateral cuneate fasciculus.

14.039 In the diagram above, ____ contains cell bodies of stretch receptors in the muscle spindles.

A is the correct answer.
All sensory receptors below the head have cell bodies in the dorsal root ganglion, be they GSA or GVA.

14.040 In the diagram above, ____ is the position of the corticospinal tract.

F is the correct answer.

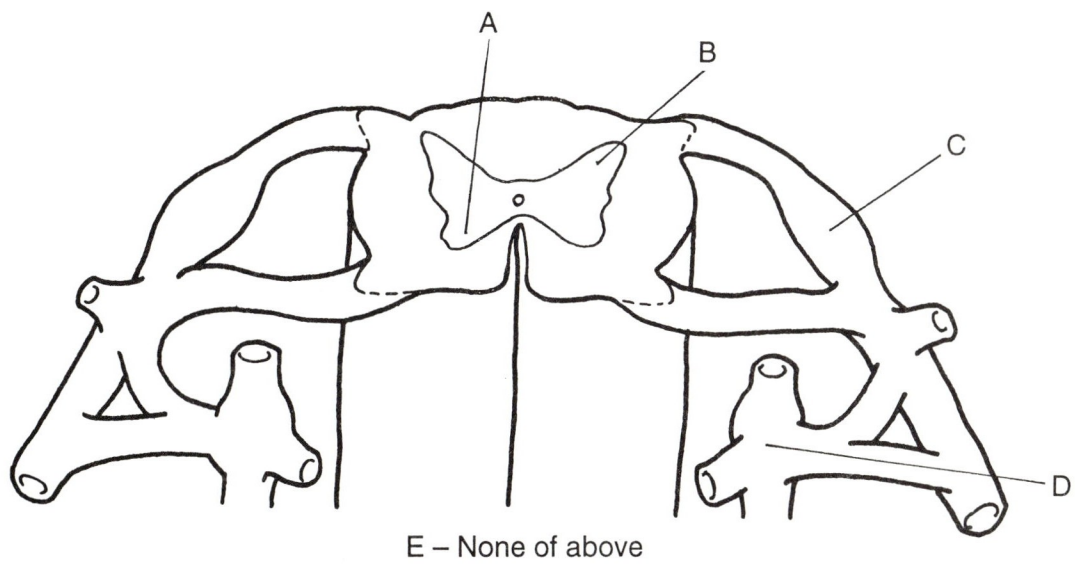

E – None of above

For questions 41 - 45, refer to the diagram above.

14.041 In the diagram above, ____ is the location of nerve cell bodies stimulating motor end plates.

A is the correct answer.
The anterior horn cells.

14.042 In the diagram above, ____ is the location of cell bodies of the greater splanchnic nerve.

E is the correct answer.
Preganglionic sympathetic cell bodies are in the lateral GVE horn.

14.043 In the diagram above, ____ is the location of cell bodies stimulating sweat gland secretions.

D is the correct answer.
Sweat glands are innervated by postganglionic cells in the sympathetic trunk.

14.044 In the diagram above, ____ is the location of preganglionic parasympathetic cell bodies.

E is the correct answer.
Parasympathetic preganglionics are in cranial nerve nuclei and S2-3-4 lateral horn area (too few to show).

14.045 In the diagram above, ____ is the location of nerve cell bodies transmitting proprioceptive impulses.

C is the correct answer.
Cell bodies of all sensory neurons lie outside of the CNS.

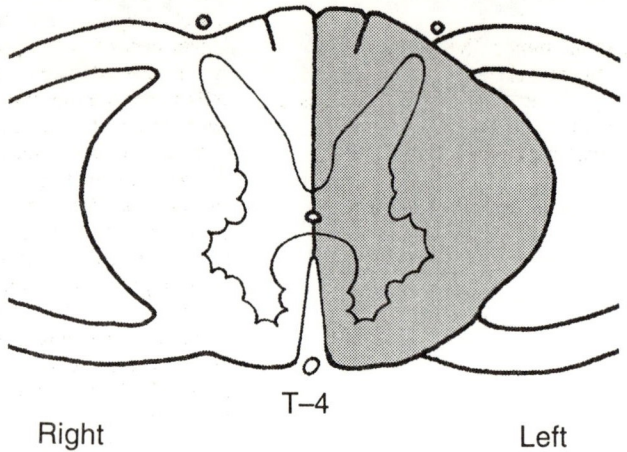

T-4

Right   Left

For questions 46 - 47, refer to the diagram above.

14.046 In the diagram above:
- A. This lesion could be produced by an acute thrombosis of the anterior spinal artery.
- B. Results in Horner's syndrome.
- C. Produces loss of pain and temperature reception on the right, below the lesion level.
- D. Results in the appearance of a plantar extensor reflex (Babinski) and spastic paralysis right leg.
- E. Results in an intention tremor of the left upper extremity.

C is the correct answer.
Crossed tracts originating on the right side will be destroyed. Uncrossed tracts originating on the left side will be destroyed. Effects will be seen below the lesion. Right gracilis, an uncrossed tract originating on the right will not be bothered. The upper extremity will not be involved in a lesion at this level. Too, an intention tremor is usually the result of injury of the cerebellum proper. Anterior spinal artery occlusion produces bilateral injury. T4 GVE neurons are too low to produce Horner's Syndrome.

14.047 Again, refer to the diagram above. One would expect to demonstrate on examination of a patient with the above defect all but one of the following:
- A. Ipsilateral spastic paralysis below the level of the lesion.
- B. Ipsilateral loss of stereognosis below the level of the lesion.
- C. Contralateral paralysis of the distal portions of the extremities.
- D. Contralateral loss of pain and temperature sensation below the lesion level.
- E. Hyperreflexia, ipsilateral.

C is the correct answer.
Paralysis below the lesion will be of the upper motor neuron type; spastic, hypertonic, hyperreflexic. Ipsilateral stereognosis will be lost below the lesion; carried in the uncrossed posterior columns. The lesion interrupts pain and temperature fibers of the spinothalamic tract which originated on the right, but crossed to the left in their ascent. The right sided motor tracts are intact - there will be no right sided motor problems.

14.048 One of the following statements is incorrect regarding muscle tone:
  A. Is controlled by gamma motor fibers innervating intrafusal fibers of the muscle spindles.
  B. Is increased to the point of spasticity with lesions of the anterior horn cells.
  C. Is regulated by afferent and efferent fibers of the gamma loop.
  D. Is the result of basic tonic firing of anterior horn cells.

B is the correct answer.
Effecting appropriate muscle tone as ordered by higher centers is the province of the gamma loop system (named from the small gamma anterior horn cells that innervate the intrafusal spindle muscle). Large muscle tone is manifest by its resistance to stretch. This is caused by tonic firing of the large alpha motor neurons of the anterior horn. The gamma loop "set" of tone is effected by gamma efferents to the intrafusal muscle. Spindle afferents recognize any deviation from this set and will order alpha motor neuron action to return the muscle to the "set" tone. As pointed out in previous comments, loss of integrating (largely inhibitory) influences from higher centers results in disinhibition of the loop, hyperactivity of the loop, hypertonicity and hyperreflexia.

14.049 Concerning the stretch reflex:
  A. Non-contractile portions of muscle spindles are supplied with stretch-sensitive afferent nerve fibers.
  B. Any active joint movement requires both excitatory and inhibitory stimuli from the spinal cord.
  C. Most reflex activity in skeletal muscle may be inhibited by activity of higher centers in the central nervous system.
  D. Intrafusal fibers cannot contract.
  E. Is important in maintaining posture.

D is the correct answer.
The essence of loop activity is the contractibility of intrafusal muscle fibers. The non-contractile portion of the spindle contains stretch-sensitive afferent nerve endings. The muscle spindle gamma loop system is most important in maintaining posture. The inhibitory potential of higher centers is in part what they are all about, i.e., control and integration. Muscle tone balances agonists and antagonists. If a joint is to move by muscle action, agonists must be stimulated and antagonists relaxed.

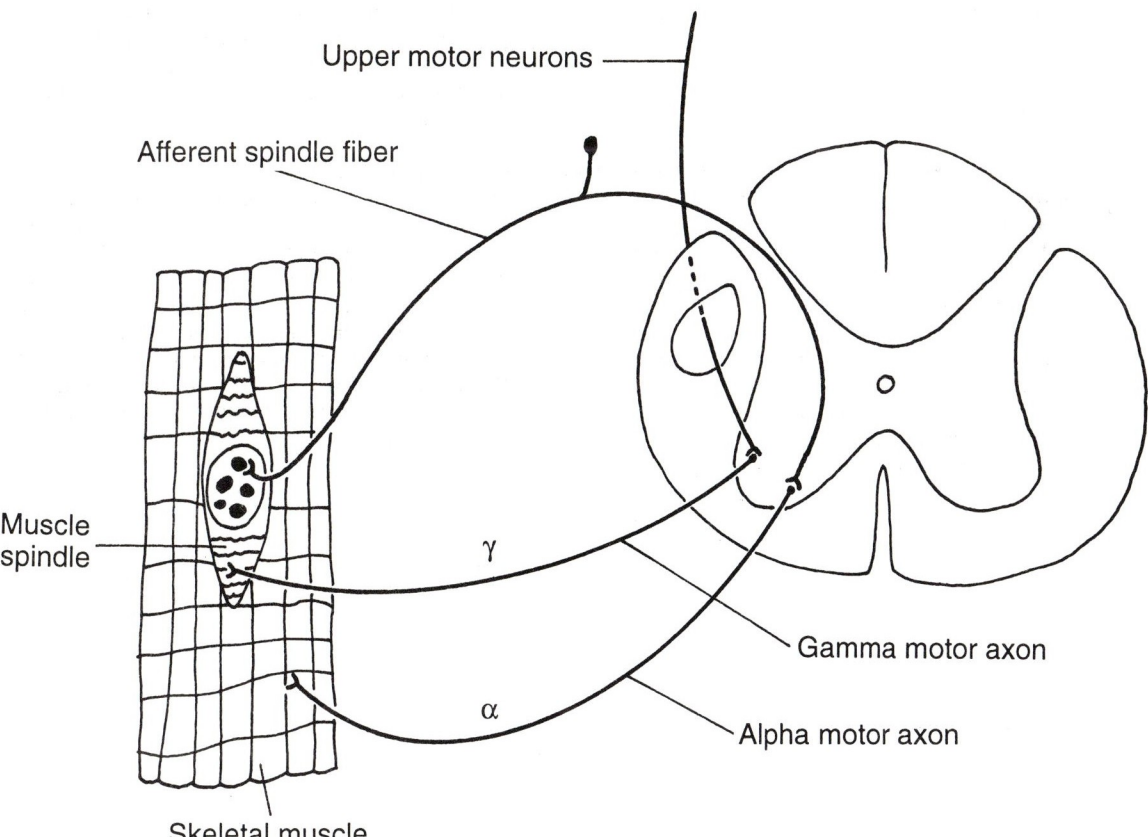

For question 50, refer to the diagram above.

14.050 One of the following statements is incorrect regarding the gamma loop:
    A. Stretch of extrafusal muscle will result in increased firing of the sensory arm of the loop.
    B. Injury to the afferent limb of the loop will produce no change in tone in the muscle supplied because the gamma motor neuron is still intact.
    C. The gamma loop is affected by collaterals from corticospinal tracts.
    D. A lesion in the posterior limb of the internal capsule will disinhibit the gamma loop.
    E. Disinhibition of the gamma motor neuron will result in increased tone and exaggerated stretch reflexes in the area supplied.

B is the correct answer.
As intra- and extrafusal muscle fibers are attached in parallel, stretch of extrafusal muscle will also stretch intrafusal muscle, and cause increased firing of the loop. As stated above, higher centers order the "set" of the loop - the appropriate intrafusal muscle fiber tension. Injury to any part of the loop itself will destroy function of the loop and result in flaccidity. A lesion in the posterior limb of the internal capsule will leave the loop intact and uncontrolled, due to destruction of the corticospinal tract. With the inhibitory lid off, the loop can charge ahead on its own. Increased tone and stretch reflex strength will result.

14.051 A patient demonstrating absence of the patellar stretch reflex:
    A. may demonstrate spasticity of the knee extensors.
    B. may have an injury or disease interrupting the fibers of higher centers which run to the anterior horn cells innervating the extensor muscles of the knee.
    C. will usually exhibit the plantar withdrawal (Babinski) reflex on the same side.
    D. may have an interruption of the dorsal roots containing the sensory nerves running from the extensor muscles of the knee joint.
    E. usually has anesthesia of the leg.

D is the correct answer.
Interruption of dorsal root fibers emanating from the knee extensors will eliminate afferent impulses from muscle spindles of these muscles. Without this sensory input, there will be no reactive reflex response to a stretch of the knee extensors (patellar tap). (A similar loss of course will result from lesions of the anterior horn cell neurons that innervate these extensor muscles.) Hypertonic spasticity also depends on an intact muscle spindle-anterior horn cell loop. Injury to fibers from higher centers running to the anterior horn cells of the knee extensors will not interfere with the gamma loop. Rather the loss of these higher influences will disinhibit the loop and the patellar reflex will be enhanced. Similarly, the primitive plantar Babinski reflex appears with loss of higher center control, and usually appears in company with an overactive patellar response. The skin of the leg is innervated by L4, 5, S1, 2. Knee extensors are innervated by L1, 2. Local lesions at the higher level would produce loss of the patellar reflex without influencing sensation of the leg.

14.052 Select the one incorrect statement. Unilateral upper motor neuron lesions result in these findings on the involved side:
    A. Release of gamma loop and hyperreflexia.
    B. Flaccid paralysis.
    C. Babinski reflex.
    D. Increase in resting muscle tone.
    E. Persistence of axial motor function.

B is the correct answer.
As in the exercise above, with loss of higher center inhibiting activity the gamma loop is uncontrolled and hyperactive. Hyperreflexia and hypertonicity result. The appearance of the Babinski reflex is also the result of a release from higher center control - largely inhibitory. Axial motor function persists; as it is bilaterally innervated by the ventral corticospinal tract.

14.053 Which of the following does not influence motor neurons of the cord either directly or with the intervention of small intercalated neurons?
    A. Vestibular nuclei.
    B. Reticular formation.
    C. Red nucleus.
    D. Corpus striatum.
    E. Midbrain tectal nuclei.

D is the correct answer.
Vestibulo-, reticulo-, rubro-, and tectospinal tracts all influence motor neurons of the cord. The corpus striatum, while important in the planning and initiation of motor function, does not directly influence the firing of anterior horn cells.

14.054 The anterior white commissure contains crossing fibers of which of the following?
- A. Fasciculus gracilis.
- B. Fasciculus cuneatus.
- C. Lateral corticospinal tract.
- D. Dorsal spinocerebellar tract.
- E. Spinothalamic tract.

E is the correct answer.
The only other decussating tract in the anterior commisure is the ventral spinocerebellar tract (which crosses back over the superior cerebellar brachium and the superior medullary velum).

14.055 The dorsal root ganglion of the fifth thoracic spinal nerve:
- A. contains nerve cell bodies of pain fibers arising in the 4th and 5th and 6th intercostal spaces.
- B. sends pain impulses to ascend in the fasciculus cuneatus of the spinal cord.
- C. lies near the intervertebral foramen between T4 and T5.
- D. transmits preganglionic sympathetic axons (GVE) to the heart.
- E. contains nerve cell bodies of stereognostic receptors. Central processes of this modality cross to the contralateral side after synapse in the cord.

A is the correct answer.
There is at lease a one segment sensory overlap in these straightforward segmental trunk nerves. Pain impulses travel in the spinothalamic tract. T5 exits between T5 and T6. Its DRG lies in the T5-T6 intervertebral foramen. Preganglionic sympathetic axons are bound for the T5 body segment and the greater splanchnic nerve. The stereognostic modality travels to the brain on the side of entrance. It crosses in the medulla to help form the medial lemniscus (cuneate nucleus and internal arcuate fibers).

14.056 Regarding spinal nerves, which one is correct?
- A. The common peroneal nerve carries fibers from the ventral primary rami of lumbar and sacral spinal nerves.
- B. The dorsal primary rami of the sacral spinal nerves carry no postganglionic fibers.
- C. The L5 spinal nerve exits the vertebral canal between the L4 and L5 vertebrae.
- D. All five sacral spinal nerves carry preganglionic parasympathetic nerve fibers.
- E. All five lumbar spinal nerves carry preganglionic sympathetic nerve fibers.

A is the correct answer.
The common peroneal nerve is derived from the sciatic plexus, and carries both lumbar and sacral fibers. All spinal nerves carry postganglionic sympathetic fibers in both dorsal and ventral primary rami. As the preganglionic sympathetic outflow stops at L2, only L1 and L2 carry preganglionic fibers. Preganglionic parasympathetic fibers classically issue only from S2, 3, 4. The L5 spinal nerve issues between L5 and the sacrum. Caudal to the cervical spine, all spinal nerves issue below the vertebra signifying the segment of origin.

14.057 Locate the one correct statement regarding the cervical spinal cord.
- A. Contributes preganglionic sympathetic fibers to the thoracic sympathetic chain.
- B. Has a prominent lateral horn.
- C. Contributes to the formation of eight cervical spinal nerves.
- D. Contains parasympathetic neurons whose axons form the cardiac nerves.

C is the correct answer.
Again, sympathetic outflow is from T1 to L2. Nerves from these segments are the only ones carrying preganglionics. Sympathetic postganglionics, derived from trunk ganglia, are carried in every spinal nerve. These segments demonstrate no GVE lateral horn. There are no parasympathetic neurons in the cervical cord. Parasympathetic cardiac nerves come from the vagi.

14.058 Which of the following structures will not be found dorsal to the sulcus limitans?
- A. Substantia gelatinosa.
- B. GVA cell column.
- C. Nucleus dorsalis (Clarke's column).
- D. Lateral gray horn.
- E. Fasciculus gracilis.

D is the correct answer.
Only sensory neurons are found dorsal to the sulcus limitans. (A-B-C-E). The lateral horn contains sympathetic GVE neurons.

14.059 Which statement is not true regarding the conus medullaris?
- A. Its caudal level is usually in the area of the intervertebral disc between L1 and L2..
- B. It contains anterior horn cells whose axons are contained in the sacral nerves.
- C. It is attached to the caudal extremity of the dural sac by the pial filum terminale.
- D. GVE cells are present.
- E. Contributes fibers to spinal nerve L1.

E is the correct answer.
While the conus exists at L1 vertebra, its anterior horn cells produce fibers for sacral nerves. Anterior horn cells producing spinal nerve L1 will exit the cord several vertebral segments higher than vertebra L1. The conus is tethered by the filum terminale. GVE cells are present in the conus providing the parasympathetic outflow in nerves S2, 3, 4.

14.060 The anterior horn of the gray matter of the cervical cord is accurately described in one statement below.
- A. Contains ascending fibers from pain receptors in the hand.
- B. Contains neurons derived from neural crest tissue.
- C. Contains motor neurons which innervate smooth muscle of the bronchi.
- D. Contains neurons which supply general somatic efferent fibers to the interossei muscles of the hand.

D is the correct answer.
Anterior horn cells supply skeletal, not smooth muscle. They will supply skeletal muscle of the hand. No sensory fibers are present in the anterior horn. The neural crest generates neurons residing outside the brain and spinal cord.

14.061 Identify the one correct statement regarding the cervical spinal cord.
- A. Contributes preganglionic sympathetic fibers to the thoracic sympathetic chain.
- B. Has a prominent fasciculus cuneatus.
- C. Is supplied by the Artery of Adamkiewicz.
- D. Is tightly wrapped by the protecting dura.
- E. Has a greater ratio of gray matter to white matter than the lumbar cord.

B is the correct answer.
Preganglionic sympathetic fibers originate in segments T1 through L2. The cervical cord is loosely wrapped by the dura matter. The artery of Adamkiewicz arises from lower segmental arteries - anywhere from T10 to L2. Much more white matter is present in the cervical cord than in the lumbar cord. Both areas have large central gray "butterflies". The fasciculus cuneatus is prominent in the cervical cord area as it contains the multitude of fibers carrying conscious proprioceptive and stereognostic information from T6 and up.

14.062 Identify the one incorrect statement regarding the development and structure of the spinal cord.
- A. The sulcus limitans divides the alar (sensory) plate from the basal (motor) plate.
- B. The neural crest tissue gives rise to all dorsal root ganglion neurons.
- C. The nucleus dorsalis extends from the level of C8 to L3.
- D. Fibers composing the fasciculus gracilis have their cell bodies located in dorsal root ganglia of T6 down, on the ipsilateral side.
- E. Fibers of the ventral corticospinal tract crossed at the level of the pyramidal decussation.

E is the correct answer.
The ventral corticospinal tract originates as the uncrossed portion of the corticospinal tract at the level of the decussation of the pyramids.

14.063 The junction of the spinal cord and medulla is most accurately marked by:
- A. upper rootlet of the second cervical nerve.
- B. upper rootlet of the first cervical nerve.
- C. level of the atlas.
- D. level of the foramen magnum.
- E. level of the caudal end of the fourth ventricle.

B is the correct answer.
The best identification of the cord-brainstem junction is the location of the upper rootlet of C1. Other landmarks listed are either incorrect (A and E), or not as accurate as B.

14.064 Identify the one incorrect statement.
A. The dorsal intermediate sulcus extends the length of the cord.
B. The lateral gray horn (intermediolateral cell column) extends from T1 through L2.
C. Dorsal roots enter the cord in the dorsal lateral fissure, and ventral roots exit via the ventral lateral fissure.
D. The nucleus dorsalis (column of Clarke) is located in the medial base of the dorsal gray horn.
E. The nucleus proprius makes up the bulk of the dorsal gray horn.

A is the correct answer.
The dorsal intermediate sulcus is the groove between gracile and cuneate fasciculi; thus only present above T6.

14.065 Choose the INCORRECT answer below. Reflex activity governed by muscle spindles and tendon organs:
A. is necessary for the performance of any skeletal muscle action.
B. is necessary for the subconscious maintenance of posture.
C. in the human, is governed by stimulatory and inhibitory directions from the brain.
D. when abnormal, may offer important clues as to the nature and location of the neurologic dysfunction.
E. will be enhanced subsequent to lesions of the anterior horn cells.

E is the correct answer.
The anterior horn cell is an integral part of the gamma loop through which the muscle spindles and tendon organs operate. Loss of anterior horn cells results in loss of activity of any motor element they supply. All the other answers are correct.

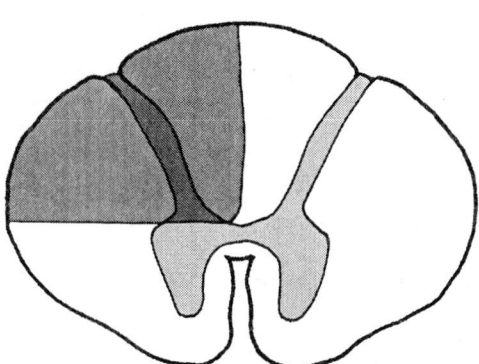

For question 66, refer to the diagram above.

14.066 Choose the CORRECT symptom corresponding to the lesion indicated by the arrow and hatching in the drawing above.
A. Lower motor neuron paralysis.
B. Loss of pain sensation contralateral to the lesion.
C. Loss of two-point discrimination contralateral to the lesion.
D. Loss of position sense ipsilateral to the lesion.
E. Intention tremor.

D is the correct answer.
The lesion destroys the posterior columns – gracilis and cuneatus. Conscious proprioception – position sense – is one of the responsibilities of the posterior columns. The fibers in these tracts do not cross until they reach the lower medulla. Thus, the loss of position sense with a lesion at this level is ipsilateral. There will be no loss of pain sensation as this modality is carried in the more anterior spinothalamic tracts. There will be ipsilateral upper motor neuron paralysis due to interruption of the lateral corticospinal tract.

112   Spinal Cord

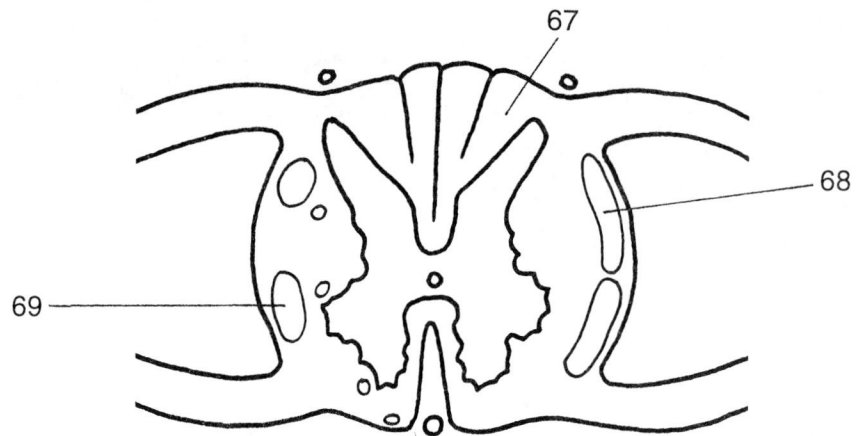

For questions 67 - 69, refer to the diagram above.

14.067 In the diagram above, a patient with a selective lesion of the pathway labeled (67) would:
A. have difficulty walking downstairs.
B. demonstrate hyperreflexia.
C. show loss of vibratory sense in the foot.
D. likely have difficulty in tying shoestrings on dark winter mornings.
E. have anterograde degeneration in the nucleus gracilis.

D is the correct answer.
The cuneate fasciculus is lesioned. It reports proprioception and stereognosis from the upper one half of the body to consciousness. Thus, its destruction would make tying shoelaces in the dark (without visual substitution for the lost cuneate function) a difficult performance. The gracilis tract is not involved; conscious proprioception remains intact in the lower one half of the body. Hyperreflexia is not a result of sensory loss.

14.068 In the diagram above, the tract labeled (68):
A. relays information of spinal cord activity to the cerebellum.
B. projects primarily to the ventral posterolateral thalamus.
C. is part of the voluntary motor system.
D. is part of the reticular system.
E. enters the superior cerebellar peduncle.

A is the correct answer.
The dorsal spinocerebellar tract. This tract reports information regarding spinal cord activity to the cerebellum, entering via the inferior cerebellar peduncle. Its fibers end in the cerebellum and do not reach the thalamus. Its information is in the unconsciousness realm and not a part of the voluntary motor system (though spinocerebellar information and feedback is necessary for the maintenance of well coordinated voluntary action). The dorsal spinocerebellar tract is not part of the reticular system.

14.069 In the diagram above, the tract labeled (69):
A. synapses upon alpha-motor neurons.
B. originates primarily from the vestibular nucleus.
C. originates from the dorsal nucleus of Clarke.
D. relays pain and temperature information.
E. None of the above.

D is the correct answer.
The spinothalamic tract. Pain and temperature sensations are its province. It has nothing to do with alpha motor neurons, the cerebellum, or the vestibular apparatus.

14.070 Which pathway(s) directly affect(s) alpha motor neurons?
A. Corticospinal tract.
B. Rubrospinal tract.
C. Tectospinal tract.
D. Reticulospinal tract.
E. All of the above.

E is the correct answer.
Many influences play upon alpha motor neurons; including all of these listed. They are all important in human motor activity, reflex and voluntary. Muscle tone, precision of movement, etc., etc., all are the result of direction from above via descending tracts like these.

14.071 Inability to maintain stable erect posture with eyes closed is most closely associated with:
   A. Cavitation of the central canal of the spinal cord destroying the anterior white commissure (syringomyelia).
   B. Hemorrhage from a central branch of the posterior cerebral artery.
   C. Hemisection of the left thoracic spinal cord (Brown-Sequard syndrome).
   D. Hemisection of the right thoracic spinal cord.
   E. Demyelinating plaque of multiple sclerosis destroying function of the posterior columns (gracilis and cuneatus).

E is the correct answer.
Vision would help. Without eyes, the dorsal columns keep us erect.

14.072 Bilateral loss of pain and temperature sensation only over the upper trunk is most closely associated with:
   A. Cavitation of the central canal of the spinal cord destroying the anterior white commissure (syringomyelia).
   B. Hemorrhage from a central branch of the posterior cerebral artery.
   C. Hemisection of the left thoracic spinal cord (Brown-Sequard syndrome).
   D. Hemisection of the right thoracic spinal cord.
   E. Demyelinating plaque of multiple sclerosis destroying function of the posterior columns (gracilis and cuneatus).

A is the correct answer.
This band-like loss must involve the spinothalamic tracts in a local lesion, as pain and temperature elsewhere are intact. Destruction of the anterior white commissure would catch crossing spinothalamic fibers from both sides at the level of the lesion. Both sides would be numbed. Syringomyelia does just this.

14.073 Loss of pain and temperature in left lower extremity is most closely associated with:
   A. Cavitation of the central canal of the spinal cord destroying the anterior white commissure (syringomyelia).
   B. Hemorrhage from a central branch of the posterior cerebral artery.
   C. Hemisection of the left thoracic spinal cord (Brown-Sequard syndrome).
   D. Hemisection of the right thoracic spinal cord.
   E. Demyelinating plaque of multiple sclerosis destroying function of the posterior columns (gracilis and cuneatus).

D is the correct answer.
Hemisection of the cord in the thoracic region will interrupt the crossed spinothalamic fibers carrying pain and temperature stimuli from the contralateral lower extremity.

14.074 Spasticity and hyperreflexia in the right lower extremity is most closely associated with:
   A. Cavitation of the central canal of the spinal cord destroying the anterior white commissure (syringomyelia).
   B. Hemorrhage from a central branch of the right posterior cerebral artery.
   C. Hemisection of the left thoracic spinal cord (Brown-Sequard syndrome).
   D. Hemisection of the right thoracic spinal cord.
   E. Demyelinating plaque of multiple sclerosis destroying function of the posterior columns (gracilis and cuneatus).

D is the correct answer.
Right cord hemisection interrupts the already crossed corticospinal fibers and all the other descending tracts playing upon the ipsilateral anterior horn cells. Thus right thoracic cord hemisection will produce right lower extremity paralysis, spasticity, and hyperreflexia - the classical upper motor neuron injury scenario, due to loss of direction and release from control of the alpha motor neurons.

SECTION 15: THALAMUS

15.001 One of the following statements regarding the thalamus is CORRECT:
   A. It contains the neurons of origin of the thalamic fasciculus.
   B. Contains a ventricle which is connected with the lateral ventricle by the foramen of Magendie.
   C. Is by-passed by the brachium of the inferior colliculus.
   D. Is heavily interconnected with the cortical sensory and motor areas.
   E. Is a product of the mesencephalic brain vesicle.

D is the correct answer.
The output of the thalamus involves all cortical areas. It is derived from the prosencephalon. It receives auditory input from the inferior colliculus, via the inferior collicular brachium, to the medial geniculate thalamic nucleus. Neurons originating the thalamic fasciculus reside in the globus pallidus and the cerebellar dentate nucleus.

15.002 Which of the following structures is/are part of the epithalamus?
   A. Caudate nucleus.
   B. Putamen.
   C. Globus pallidus.
   D. Habenula.
   E. Stria terminalis.

D is the correct answer.
Do not confuse the stria terminalis with the stria medullaris thalami. The former is a tract connecting the amygdala with the hypothalamus and the septal nuclei.

15.003 Which of the following structures is not part of the subthalamus?
   A. Subthalamic nucleus of Luys.
   B. Rostral portion of the substantia nigra.
   C. Rostral red nucleus.
   D. Pre-rubral field.
   E. Pineal body.

E is the correct answer.
The pineal body is an epithalamic structure.

15.004 Choose the one CORRECT statement regarding the thalamus.
   A. All sensory information save olfactory is relayed through the thalamus.
   B. The pulvinar has afferent input from sensory receptors in addition to the to-and-fro communications with the cerebral cortex.
   C. Is connected with the limbic system through the anterior commissure.
   D. Has no significant connection with the motor system.
   E. The thalamic VPL nucleus has important connections with the cerebellum.

A is the correct answer.
No significant direct sensory input to the pulvinar is recognized. The anterior commissure has little to do with the thalamus in performing its job of connecting the temporal lobes and the olfactory tracts. The cerebellum is connected to the VL - not the VPL nucleus. The thalamus has wide connections with the motor system (dentato-thalamic, pallido-thalamic, etc., etc.)

15.005 Regarding the dorsal thalamus:
   A. Contains nuclei and fibers of the reticular system.
   B. Is an important actor in motor system activity planning and execution.
   C. Is nourished by the aqueduct of Sylvius.
   D. Furnishes the majority of the fibers in the internal capsule.
   E. Projects to all areas of the cortex.

C is the correct answer.
The aqueduct runs through the midbrain. All the other statements are correct.

15.006 Which of the following is not a specific thalamic relay nucleus?
- A. Ventral lateral (VL).
- B. Medial geniculate.
- C. Pulvinar.
- D. Nucleus ventralis posterolateralis (VPL).
- E. Nucleus ventralis posteromedialis (VPM).

C is the correct answer.
Specific nuclei of the thalamus are reciprocally connected with specific areas of the cerebral cortex. The pulvinar is the only one listed that does not fulfill this description. It projects to a rather diffuse area of occipital cortex and neighboring parietal and temporal areas. VL - precentral gyrus, medial geniculate - transverse temporal gyrus, VPL and VPM - postcentral gyrus are reciprocally connected pairs.

15.007 Regarding the fibers of the stria medullaris thalami, select the INCORRECT statement:
- A. originate in the parahippocampal gyrus.
- B. connect the septal area with the habenular nuclei.
- C. are considered part of the epithalamus.
- D. are in close relation to the third ventricle and the dorsolateral surface of the dorsal thalamus.
- E. information carried is relayed to the midbrain reticular system center.

A is the correct answer.
Fibers of the stria medullaris thalami originate mainly in the septal area nuclei. All the other statements are accurate.

15.008 Which of the following is not a thalamic "association" nucleus?
- A. Dorsomedial nucleus (DM).
- B. Nucleus lateralis posterior (LP).
- C. Nucleus lateralis dorsalis (LD).
- D. Pulvinar.
- E. Nucleus ventralis posteromedialis.

E is the correct answer.
The nucleus ventralis posteromedialis (VPM) is a specific nucleus relaying sensory stimuli from the head to the postcentral gyrus.

15.009 Select the one INCORRECT statement concerning the subthalamus.
- A. In an appropriate coronal section, the subthalamus can be seen lateral to the internal capsule.
- B. Histologically, it is quite similar to the globus pallidus.
- C. It contains an important "extra-pyramidal" nucleus.
- D. The globus pallidus provides the principle input to the subthalamus.
- E. A lesion in the subthalamus, usually vascular in origin, produces hemiballismus.

A is the correct answer.
The subthalamus and the globus pallidus are quite similar histologically. It is as if they were split apart by the development of the internal capsule. They are still closely connected. The important nucleus here is the subthalamic nucleus. Injury to this nucleus results in the appearance of hemiballismus. The subthalamus is medial to the internal capsule.

15.010 Are all the following statements regarding the prerubral field correct?
- A. Contains important contributions from the basal ganglia and the neocerebellum.
- B. Is responsible for the pupillary light reflex.
- C. If damaged, serious motor dyskinesias appear.
- D. Contains the thalamic fasciculus which communicates with the VL and VA nuclei of the dorsal thalamus.
- E. Is entered by both the ansa lenticularis and the lenticular fasciculus.

B is the correct answer.
The prerubral field is not concerned with the light reflex which is handled in the pretectal area, a bit further caudally. The prerubral field is the area of coming together of many important tracts bound for the thalamus. These include the dentato-thalamic tract, ansa lenticularis, and lenticular fasciculus. (Also, the main sensory tracts heading for the VPL/VPM muscle their way through this area.) A very busy place! The motor oriented tracts from the cerebellum and globus pallidus unite here to form the thalamic fasciculus which ascends to the VA/VL thalamic nuclei.

15.011 The lenticular fasciculus consists of fibers which:
- A. arise in the globus pallidus of the corpus striatum.
- B. pass through the internal capsule to reach the subthalamus.
- C. terminate in the VA and VL nuclei of the thalamus.
- D. join with the fibers of the ansa lenticularis in the prerubral area to form the thalamic fasciculus.
- E. contains contributions from the red nucleus.

E is the correct answer.
The lenticular fasciculus is "superior" to the red nucleus in the motor system hierarchy and does not depend on it for any help. All the other statements are correct.

15.012 Which one of the following thalamic relay circuits is INCORRECT?
- A. Globus pallidus - ventral anterior nucleus (VA) - premotor cortex.
- B. Mammillary body - anterior nucleus - cingulate gyrus.
- C. Superior colliculus - ventral posteromedial nucleus (VPM) - prefrontal cortex.
- D. Cerebellar dentate nucleus - ventrolateral nucleus (VL) - motor cortex.
- E. Retina - lateral geniculate - calcarine cortex.

C is the correct answer.
The superior colliculi serve as reflex centers. They direct the head and eyes in response to visual, auditory, somatic stimuli. Thus, for instance, we are able to track moving objects. Reflex movement of head and eyes keep the moving object in sharp focus. We respond reflexly with head and eyes to sounds and/or proprioceptive stimuli, etc. No significant functional relay to the VPM or prefrontal cortex is known. All the other pathways are correct.

15.013 Which association involving a thalamic nucleus is INCORRECT?
- A. Ventral posterolateral nucleus - postcentral gyrus.
- B. Lateral geniculate body - cingulate gyrus.
- C. Anterior nucleus - mammillary body.
- D. Ventral anterior nucleus - globus pallidus.
- E. Ventral lateral nucleus - brachium conjunctivum.

B is the correct answer.
The lateral geniculate might be paired with the optic tract or the visual cortex, but never with the limbic cingulate gyrus. All the other pairings make sense.

15.014 The hypothalamus is directly concerned with all of the following EXCEPT:
- A. water balance.
- B. appetite.
- C. olfaction.
- D. endocrine function.
- E. visceral correlates of emotion.

C is the correct answer.
Olfaction may concern the hypothalamus in that certain odors may be relayed to it as having high emotional content, to which the hypothalamus must respond; e.g. the flow of digestive juices ordered by the hypothalamus through the autonomic system when a starving man detects the odor of food. However, this kind of connection is not nearly as direct as the others listed.

15.015 Regarding the hypothalamus, all of the following statements are correct EXCEPT:
- A. Receives major limbic input via the parahippocampal gyrus and hippocampal formation.
- B. Has reciprocal connections with midbrain tegmentum.
- C. Exerts influence upon the cingulate cortex via thalamic nuclei.
- D. Is needed for coordinated movements.
- E. Has reciprocal connections with the amygdala.

D is the correct answer.
The hypothalamus is not necessary for the performance of coordinated movements. The hypothalamus is strongly connected to the hippocampus and amygdala via the fornix, and the stria terminalis and amygdalofugal tracts of the amygdala. The hypothalamic mammillary thalamic tract to the anterior nucleus of the thalamus affords strong hypothalamic influence on the cingulate gyrus through the many anterior nucleus-cingulate gyrus connections. Connections with the midbrain allow hypothalamic control of reticular system autonomic nervous system nuclei located in the midbrain tegmentum. All these connections are two-way streets.

SECTION 16: TRACTS

16.001 Regarding the dorsal spinocerebellar tract, all the following statements are CORRECT except:
A. Serves unconscious proprioception.
B. Is a crossed tract.
C. Receives input from muscle spindles and tendon organs.
D. Reaches the cerebellum through the inferior cerebellar peduncle.
E. Communicates with the paleocerebellar vermis.

B is the correct answer.
The dorsal spinocerebellar tract does not cross. It ascends to the cerebellum via the inferior cerebellar peduncle. The ventral spinocerebellar tract does cross in the cord, ascends contralaterally to enter the superior cerebellar peduncle whence it crosses back to the side of origin via the superior medullary velum to enter the cerebellum. All the other statements are correct.

16.002 Concerning the corticospinal tracts, one statement below is CORRECT:
A. Arise from large Purkinje cells in the precentral gyrus.
B. Descend through the posterior limb of the internal capsule.
C. Undergo 90% decussation in the lower pons.
D. Occupy a position in the dorsal fasciculus of the spinal cord.
E. May be categorized as "lower motor neuron" fibers.

B is the correct answer.
These fibers arise from large Betz cells in the precentral gyrus, descend in the posterior limb of the internal capsule, decussate in the lower medulla, and are found in the dorsal part of the lateral fasciculus. By definition, fibers stimulating anterior horn cells from above are termed "upper motor" neuron fibers. The anterior horn cell is the "lower motor" neuron.

16.003 Describing the dorsolateral fasciculus of Lissauer, one statement below is INCORRECT:
A. Lies between the apex of the dorsal horn and the exterior surface of the spinal cord.
B. Consists of ascending and descending fibers from the medial processes of dorsal root ganglion neurons.
C. Branches from fibers in Lissauer's tract carrying nociceptive stimuli enter the dorsal horn substantia gelatinosa.
D. Progresses to the thalamus as the dorsolateral lemniscus.
E. Serves to spread incoming pain and temperature stimuli among adjacent spinal cord segments.

D is the correct answer.
There is no such structure as the "dorsolateral lemniscus" - no matter how good and impressive it sounds! The other statements accurately depict the dorsolateral fasciculus.

16.004 Regarding the brachium conjunctivum, the following statements are CORRECT except:
A. Decussates upon entering the midbrain.
B. Contains output from cerebellar Purkinje cells.
C. Anatomically forms the walls of the upper fourth ventricle.
D. Enters the prerubral field of the diencephalic subthalamus.
E. Furnishes the main avenue of communication between the neocerebellum and the thalamus-motor cortex.

B is the correct answer.
Purkinje cells project to the cerebellar roof nuclei. Their axons do not leave the cerebellum. The brachium conjunctivum carries fibers from the cerebellar roof nuclei; by far the largest volume from the dentate nucleus. All the other statements are correct.

16.005 Which of the following nuclei is situated in the medulla?
A. Trochlear.
B. Hypoglossal.
C. Facial.
D. Motor nucleus of trigeminal.
E. Nucleus of the lateral lemniscus.

B is the correct answer.

16.006 Select the discordant pair:
A. Dorsal spinocerebellar tract - spinocerebellum.
B. Medial lemniscus - internal arcuate fibers.
C. Unconscious proprioception - ventral posteromedial nucleus thalamus.
D. Tractus solitarius - visceral sensation.
E. Hypoglossal nerve - general somatic efferent.

C is the correct answer.
Fibers carrying the modality of unconscious proprioception reside in the spinocerebellar tracts which end in the cerebellum. The VPM thalamus is a switchboard-relay point for sensory modalities from the head bound for the somatosensory cortex and conscious recognition.

16.007 A lesion destroying the crus cerebri on the right would produce:
A. flaccid paralysis of the muscles of the left arm and leg.
B. inability to close the right eye tightly.
C. weakness of the left side of the tongue.
D. loss of hearing in the left ear.
E. loss of phonation.

C is the correct answer.
There is a variable amount of crossing of corticobulbar fibers carried in the cerebral peduncle. Those to the hypoglossal nucleus are largely crossed fibers. Thus tongue weakness contralateral to the peduncular injury is to be expected. There is sufficient bilaterality of innervation of the upper face and larynx to permit normal function of these areas despite a unilateral peduncular lesion. Upper motor lesions disinhibit the gamma loop mechanism, with resulting overactivity and spasticity.

16.008 Which statement is INCORRECT regarding the spinothalamic tract?
A. Cell bodies are in the substantia gelatinosa.
B. Component is GSA.
C. Runs in the anterolateral white matter of the cord and stem tegmentum.
D. Terminates in the ventral posterolateral nucleus of the thalamus.
E. Crosses in the midbrain on the way to higher centers.

E is the correct answer.
The GSA spinothalamic tract crosses in the spinal cord as it arises from neurons in the substantia gelatinosa. From its crossed anterolateral position in the cord it proceeds up into the thalamus in the same anterolateral position throughout the brain stem.

16.009 Left hemisection of the spinal cord at C4 would interrupt all the following except:
A. Lateral corticospinal fibers from the right hemisphere.
B. Spinothalamic fibers carrying pain and thermal sense from the right leg.
C. Fasciculus gracilis carrying vibration sense from the right leg.
D. Posterior spinocerebellar input from the left side of the body.
E. Ventral corticospinal fibers from the left hemisphere.

C is the correct answer.
Knowledge of the anatomy of these tracts permits easy prognostication of the results of unilateral cord injury. Lateral corticospinal fibers are already crossed by the time they enter the spinal cord. Thus hemisection effects will be manifest on the same side as the lesion. Ventral corticospinal fibers have not crossed. Fibers from the ipsilateral cortex are cut. Ascending spinothalamic fibers have already crossed. Thus loss of their function will be seen on the side contralateral to the lesion - the side from which they originated. Gracilis fibers do not cross until they reach the medulla. Thus a left sided injury in the cord will interrupt left sided gracilis (and cuneatus) fibers. The posterior spinocerebellar fibers are also uncrossed and will suffer similar ipsilateral function loss.

16.010 Find the one INCORRECT statement regarding the MLF.
  A. Is present in the midbrain.
  B. Extends to the cerebral cortex via the anterior limb of the internal capsule.
  C. Serves to connect the vestibular apparatus, the extraocular musculature, and the cerebral cortex in the management of conjugate gaze.
  D. Mediates ocular adduction in lateral gaze.
  E. If destroyed on the left side cranial to the PPRF, but caudal to the oculomotor nucleus, adduction of the left eye on attempts at lateral gaze to the right will not occur, even though convergence function in the left eye is normal.

B is the correct answer.
The MLF yokes the vestibular nuclei, nuclei of extraocular muscles, and the anterior horn cells powering the neck muscles, to enable appropriate conjugate eye movements, and head-eye positional reflexes. The MLF extends from the midbrain down through the cervical cord. In the pons at the abducens nucleus level, the reticular lateral gaze center (Paramedian pontine reticular formation - PPRF), using the MLF, engineers conjugate adduction of one eye and abduction of the other. If the left MLF is destroyed, adduction of the left eye on right lateral gaze is abolished, even though the oculomotor nerve function is otherwise intact and quite able to adduct the balky eye in convergence.

16.011 There is one CORRECT answer in the following statements regarding the corticopontine fibers:
  A. Part of motor programming process.
  B. Issue from motor areas of cortex only.
  C. Synapse with neurons of the dentate nucleus.
  D. Appear in the medulla as corticospinal fibers.
  E. Play an important role in maintenance of muscle tone.

A is the correct answer.
The corticopontine tracts contain fibers from all parts of the cortex. By synapsing with pontine nuclei in the brachium pontis, the corticopontines end. Pontine nuclear axons project to the contralateral neo-cerebellum. Output of the neo-cerebellum is to the dentate nucleus which projects it decisions back to the motor cortex via the brachium conjunctivum, thalamus, and internal capsule. This "loop" plays an important role in motor programming. Direct activation of muscle is not within the purview of the corticopontines.

16.012 Regarding the corticopontine tract, one statement below is CORRECT.
  A. Fibers in the tract make up the bulk of the cerebral peduncle.
  B. Its fibers synapse with neurons of the brainstem cranial nerve motor nuclei.
  C. It decussates in the midbrain.
  D. It is a purely sensory tract.
  E. It has massive connections with the inferior cerebellar peduncle.

A is the correct answer.
The only correct statement is A. This is a huge output from the cerebral cortex. The corticospinal and corticobulbar fibers make up only a small fraction of the bulk of the cerebral peduncles. Corticopontine fibers synapse with ipsilateral pontine nuclei. They are in the motor mode. They have nothing to do with the inferior cerebellar peduncle.

16.013 Concerning the nucleus of the spinal tract of the trigeminal nerve, all the following statements are TRUE except:
  A. Occupies the same relative position in the brainstem tegmentum as does the substantia gelatinosa in the spinal cord.
  B. Extends from the pons to the upper cervical segments of the cord.
  C. Projects to the ipsilateral VPM.
  D. Is connected with neurons in the trigeminal (semilunar) ganglion.
  E. Serves modalities of pain and temperature.

C is the correct answer.
Similar to pain-temperature tracts in the spinal cord, the trigeminal spinal tract nuclei produce axons making up the trigemino-thalamic tract which also crosses to the contralateral side in its progress to the thalamus. The trigeminal spinal tract itself has originating cell bodies in the semilunar ganglion. The nucleus of the trigeminal spinal tract and the nuclei in the substantia gelatinosa are similar in their anatomy and function.

16.014 Which of the following statements regarding the nucleus of the spinal tract of the trigeminal nerve is CORRECT?
  A. It extends cranially as the mesencephalic nucleus.
  B. It is confined to the medulla.
  C. It is so-called because it sends fibers to the spinal cord nuclei.
  D. Axons of the spinal V nucleus are distributed to the thalamus as fibers in the spinal tract of V.
  E. It has characteristics shared with the dorsal gray horn of the spinal cord, especially with the substantia gelatinosa.

E is the correct answer.
The nucleus extends from the area of the principle sensory nucleus of the trigeminal, through the lower pons, medulla, and into the upper cervical cord where it merges with the substantia gelatinosa. Its axons form the trigeminothalamic tract running to the VPM of the thalamus.

16.015 Select the one INCORRECT statement. Fibers carrying the modality of unconscious proprioception:
  A. carry impulses to the VL of the thalamus.
  B. are carried in the dorsal spinocerebellar tract which originates in the nucleus dorsalis of Clarke and the accessory cuneate nucleus.
  C. end in mossy fibers in the cerebellum.
  D. receive stimuli from muscle spindles and tendon organs.
  E. are carried in the ventral spinocerebellar tract running in the side of the spinal cord opposite to the side of origin of the impulse.

A is the correct answer.
Fibers of unconscious proprioception from muscle spindles and tendon organs, and carried in the dorsal and ventral spinocerebellar tracts, enter and end in the spinocerebellum as mossy fibers. The ventral spinocerebellar tract is crossed.

16.016 Select the one INCORRECT statement. Recognizing the position of the right arm in space requires the integrity of the:
  A. nucleus of the cuneate fasciculus, right.
  B. medial lemniscus, left.
  C. MLF.
  D. VPL of thalamus.
  E. postcentral gyrus.

C is the correct answer.
The MLF is not involved in conscious proprioception. All the other statements correctly describe parts of the conscious proprioception path to the somatosensory cortex.

16.017 Select the one CORRECT statement. A patient with loss of pain and temperature sensation over the right face and left body has a lesion in the:
  A. postcentral gyrus, right.
  B. ventral posterolateral and ventral posteromedial (VPL and VPM) nuclei of the left thalamus.
  C. brachium pontis.
  D. lateral tegmentum of the medulla, right.
  E. medial lemniscus, left.

D is the correct answer.
The crossed spinothalamic tract from the left body, and the uncrossed right tract and nucleus of the trigeminal are juxtaposed in the right lateral tegmentum of the medulla. These findings are part of a concatenation of signs and symptoms found in cases of occlusion of the posterior inferior cerebellar artery (PICA).

16.018 One of the following statements regarding the posterior columns is CORRECT:
  A. Receive their fibers from neurons of dorsal horn of the spinal cord.
  B. Carry sensory information from endings in the joints.
  C. Injury results in contralateral sensory losses.
  D. Their intracranial fibers extend into the restiform body.
  E. Their fibers terminate in the contralateral gracile and cuneate nuclei.

B is the correct answer.
Among fibers in the dorsal columns, reporting position sense and stereognosis to consciousness, there will certainly be fibers carrying sensory information from joints. Cell bodies are in dorsal root ganglia. Central processes ascend without interruption to ipsilateral nuclei in the medulla. There is no passage to the cerebellum. Injury to the tracts will cause ipsilateral symptoms.

16.019 Which of the following tracts will terminate in the nucleus ventralis posterior or medialis (VPL or VPM) of the thalamus?
A. Trigeminothalamic.
B. Lateral or auditory lemniscus.
C. Corticopontine.
D. Spinotectal.
E. Lenticular fasciculus.

A is the correct answer.
VPL and VPM are thalamic nuclei specifically charged with receiving fibers carrying cranial and bodily sensory stimuli, and relaying these stimuli to consciousness in the cortex. The trigeminothalamic tract is included. Auditory thalamic connection is the medial geniculate. The other tracts are of the motor persuasion.

16.020 Find the one INCORRECT statement regarding the lateral vestibulospinal tract.
A. Extends throughout the spinal cord.
B. Is primarily concerned with the maintenance of balance.
C. Functions in an "unconscious" reflex pattern.
D. Is found in the posterior funiculus of the cervical spinal cord.
E. May influence muscle tone.

D is the correct answer.
The lateral vestibulospinal tract effects all the functions stated. It extends throughout the spinal cord in the anterior funiculus of the cord.

16.021 Fibers of the medial longitudinal fasciculus synapse in all but one of the following nuclei:
A. Abducens.
B. Oculomotor.
C. Trochlear.
D. Facial.
E. C2 anterior horn.

D is the correct answer.
Many of the cranially directed fibers in the MLF are responsible for organizing conjugate eye movements. Obviously, synapse in eye muscle nuclei is required. Descending MLF fibers reach neck muscles which control head position in connection with eye actions.

16.022 The ability to identify the value of unseen coins in your pocket depends on the integrity of:
A. medial lemniscus.
B. nucleus dorsalis.
C. dorsal spinocerebellar tract.
D. VL of thalamus.
E. crus cerebri.

A is the correct answer.
Stereognosis is a prime function of the dorsal column-medial lemniscus-internal capsule-sensory cortex pathway. The spinocerebellar systems are not directly involved; nor is the VL thalamus or crus cerebri.

16.023 Which one of the following cannot influence the firing of a lower motor neuron?
A. Tectospinal tract.
B. Corticopontine tract.
C. Vestibulospinal tract.
D. Rubrospinal tract.
E. Corticobulbar tract.

B is the correct answer.
There are many tracts directly influencing the firing of anterior horn cells or cranial nerve motor neurons. The corticopontine tract, however, is not one of them.

16.024 The restiform body (inferior cerebellar peduncle) is composed of all but one of the following:
A. Dorsal spinocerebellar tract.
B. Olivocerebellar tract.
C. Cuneocerebellar tract.
D. External arcuate fibers.
E. Dentatothalamic tract.

E is the correct answer.
The dentatothalamic tract projects to the thalamus via the superior cerebellar peduncle. All the others traverse the inferior peduncle.

16.025 Which of the following statements regarding the pyramidal tract is INCORRECT?
A. The cell bodies are located in the precentral gyrus (Area 4 Brodmann).
B. The upper motor neuron axons synapse in the anterior horn of the spinal cord.
C. The large pyramidal cells (of Betz) are located in the precentral gyrus.
D. All fibers composing the ventral corticospinal tract decussate by crossing in the anterior white commissure.
E. The lateral corticospinal tract occupies the contralateral side of the spinal cord from the side of origin.

D is the correct answer.
Each ventral corticospinal tract innervates both sides of the axial musculature.

16.026 Which of the following statements concerning the corticobulbar tract is INCORRECT?
A. Interruption of this tract produces voluntary weakness of the contralateral facial muscles.
B. It originates in the medial one-third ("face area") of the precentral gyrus.
C. It influences all motor neurons associated with cranial nerves. (Only I, II, and VIII being purely sensory nerves do not receive corticobulbar fibers.)
D. All of its fibers cross the midline in the brainstem, and end in motor nuclei of cranial nerves on the side opposite the hemisphere of origin.
E. It runs in the midportion of the crus cerebri.

D is the correct answer.
Corticobulbar innervation of cranial nerve motor nuclei varies in degree of bilaterality. Eye muscle nerve nuclei, upper facial nucleus, and the nuclei of IX, X, XI are bilaterally innervated so that injury to one corticobulbar tract will produce no discernible loss of function. The motor nuclei of V and XII are predominantly innervated by crossed corticobulbar fibers. Some ipsilateral innervation is also present. The injury of corticobulbar fibers to V and XII produces noticeable diminution of function and atrophy contralaterally, but not a total loss. The lower facial nucleus depends entirely on crossed corticobulbar innervation. Corticobulbar loss results in facial paralysis only below the eyes. All other statements are correct.

16.027 An acute lesion of the lateral corticospinal tract in the cervical cord will cause all but one of the following:
A. flaccid paralysis.
B. exaggerated tendon reflexes and increased muscle tone.
C. late muscle atrophy.
D. spastic hemiparesis with less proximal (trunk, hip, shoulder) weakness than distal (hand, foot) weakness.
E. plantar extension response.

A is the correct answer.
This is an "upper motor neuron" paralysis; characteristically a spastic paralysis. Spasticity, increased tone, and hyperreflexia are due to disinhibition of the gamma loop. Without upper motor neuron control, the loop "goes wild". Late muscle atrophy is the result of disuse, not denervation. The primitive plantar extension response appears with loss of controlling suppressing corticospinal influence. The corticospinal tract developed to handle the intricate musculature of the distal extremities. Its loss will be most evident in diminution of function of the hands and feet. The older and more primitive axial muscles will continue to function fairly well.

16.028 The decussation of the medial lemniscus (internal arcuate fibers) is located in the:
A. spinal cord.
B. medulla.
C. pons.
D. midbrain.
E. diencephalon.

B is the correct answer.

16.029 Lateral corticospinal fibers will be found in which one of the following locations?
A. Anterior limb of the internal capsule.
B. Dorsal columns of the cord.
C. The medial one-fifth of the basis pedunculi.
D. Longitudinal fibers in the brachium pontis.
E. Anterior spinal cord funiculus.

D is the correct answer.
Corticospinal fibers are found in the posterior limb of the internal capsule. They pass through the brachium pontis to get from the midportion of the cerebral peduncle to the pyramidal ridges of the medulla. In the cord, they are found in the dorsolateral area of the lateral funiculus.

16.030 Select the one CORRECT statement: A patient with loss of pain and temperature sensation over the right face and over the left body has a lesion in:
A. somato-sensory (postcentral gyrus) right.
B. left VPL-VPM thalamus.
C. lateral tegmentum upper right medulla.
D. brachium pontis.
E. left medial lemniscus.

C is the correct answer.
In this area we find the spinal tract of the trigeminal carrying descending uncrossed pain and temperature sensation from the ipsilateral face. Pain and temperature fibers from the remainder of the body have already crossed and are ascending in the lateral medullary tegmentum. Thus a lesion here on the right could produce symptoms in the right face and the left body. Such a territorial split would not occur in areas where both tracts have crossed (A and B). The brachium pontis and medial lemniscus are not concerned with pain and temperature.

16.031 The dorsal spinocerebellar tract is most closely associated with:
A. MLF.
B. Lateral lemniscus.
C. Inferior cerebellar peduncle.
D. Medial lemniscus.
E. Spinal tract of trigeminal nerve.

C is the correct answer.

16.032 Which of the following is found in cervical cord and brainstem immediately ventral to central canal and ventricular system, in paramedian position?
A. MLF.
B. Lateral lemniscus.
C. Inferior cerebellar peduncle.
D. Medial lemniscus.
E. Spinal tract of trigeminal nerve.

A is the correct answer.

16.033 Nucleus dorsalis (Clarke) is most closely associated with:
A. MLF.
B. Lateral lemniscus.
C. Inferior cerebellar peduncle.
D. Medial lemniscus.
E. Spinal tract of trigeminal nerve.

C is the correct answer.

16.034 Which of the following is a somatotopically arranged crossed tract?
A. MLF.
B. Lateral lemniscus.
C. Inferior cerebellar peduncle.
D. Medial lemniscus.
E. Spinal tract of trigeminal nerve.

D is the correct answer.

16.035 Which of the following contains crossed and uncrossed fibers that terminate in the caudal midbrain?
A. MLF.
B. Lateral lemniscus.
C. Inferior cerebellar peduncle.
D. Medial lemniscus.
E. Spinal tract of trigeminal nerve.

B is the correct answer.

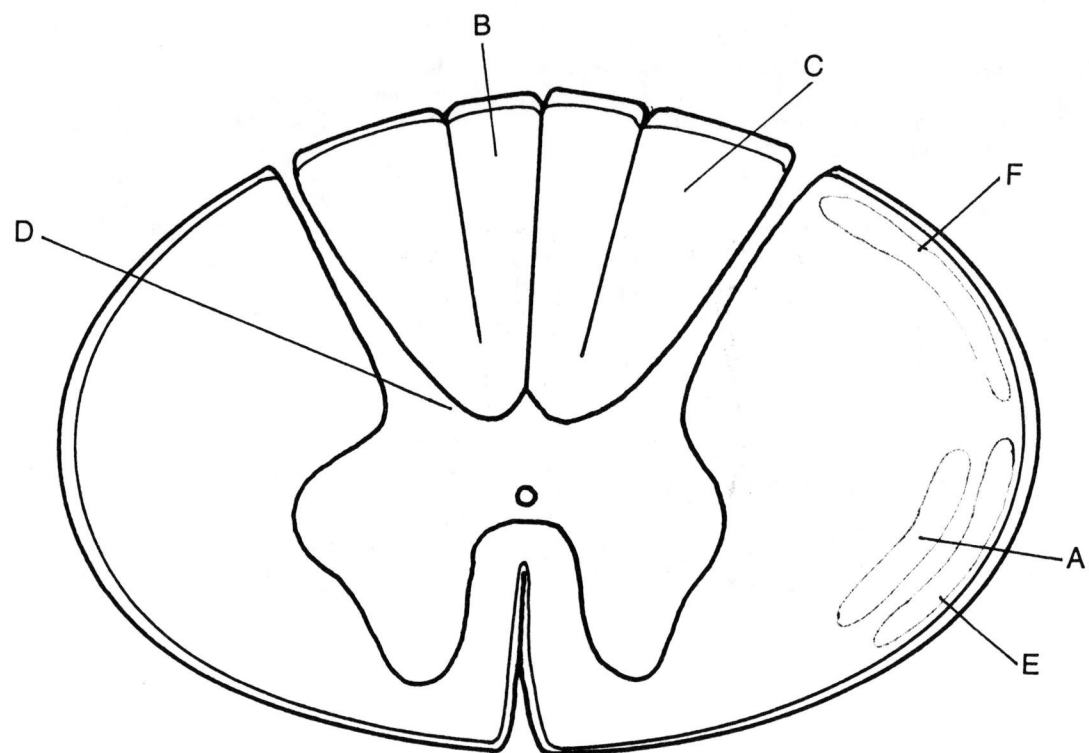

For questions 36 - 41, refer to the diagram above.

16.036 In the diagram above, which travels to the cerebellum via the restiform body?

F is the correct answer.
Dorsal spinocerebellar tract.

16.037 In the diagram above, which is associated with pain and temperature?

A is the correct answer.
Spinothalamic tract.

16.038 In the diagram above, which is associated with dorsal root ganglia of T6-down?

B is the correct answer.
Fasciculus gracilis.

16.039 In the diagram above, which reaches the cerebellum via the brachium conjunctivum?

E is the correct answer.
Ventral spinocerebellar tract.

16.040 In the diagram above, _____ is the nucleus dorsalis (Clarke's column).

D is the correct answer.

16.041 In the diagram above, which is associated with epicritic touch from upper half of body?

C is the correct answer.
Fasciculus cuneatus.

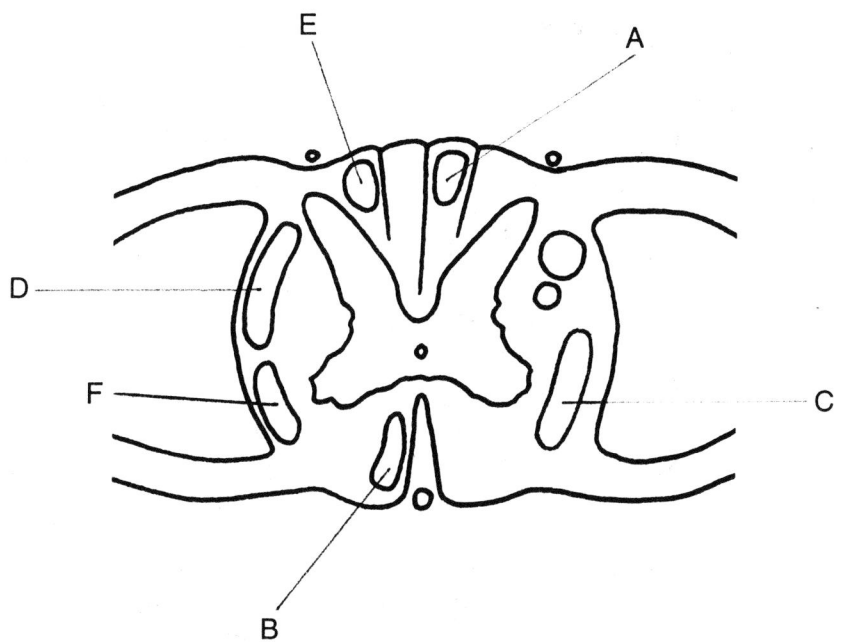

For questions 42 - 45, refer to the diagram above.

16.042 In the diagram above, fibers in tract ____ will synapse with anterior horn cells.

B is the correct answer.
Ventral spinothalamic tract.

16.043 In the diagram above, tract ____ contains stereognostic information from the lower extremity.

A is the correct answer.
Fasciculus gracilis.

16.044 In the diagram above, fibers in tract ____ have synapsed with a spinal cord nucleus of this same side.

D is the correct answer.
Dorsal spinocerebellar tract.

16.045 In the diagram above, destruction of tract ____ will result in a loss of pain sensation on the opposite side below the level of destruction.

C is the correct answer.
Spinothalamic tract.

16.046 Which of the following receives general somatic afferent fibers?
  A. solitary tract and nucleus only.
  B. trigeminal spinal tract and nucleus only.
  C. both the solitary tract/nucleus and trigeminal spinal tract/nucleus.
  D. neither solitary tract/nucleus nor trigeminal spinal tract/nucleus.

B is the correct answer.

16.047 Which of the following is involved with visceral afferent fibers?
  A. solitary tract and nucleus only.
  B. trigeminal spinal tract and nucleus only.
  C. both the solitary tract/nucleus and trigeminal spinal tract/nucleus.
  D. neither solitary tract/nucleus nor trigeminal spinal tract/nucleus.

A is the correct answer.

16.048 Which of the following projects motor fibers to muscles of mastication?
- A. solitary tract and nucleus only.
- B. trigeminal spinal tract and nucleus only.
- C. both the solitary tract/nucleus and trigeminal spinal tract/nucleus.
- D. neither solitary tract/nucleus nor trigeminal spinal tract/nucleus.

D is the correct answer.

16.049 Which of the following receives special visceral afferent fibers?
- A. solitary tract and nucleus only.
- B. trigeminal spinal tract and nucleus only.
- C. both the solitary tract/nucleus and trigeminal spinal tract/nucleus.
- D. neither solitary tract/nucleus nor trigeminal spinal tract/nucleus.

A is the correct answer.
The solitary tract receives all general visceral afferent fibers from the thoracic and abdominal viscera, and from the head. Special visceral afferents from the taste buds project to the upper end of the solitary tract and its nucleus here - the so-called gustatory nucleus. Only SVA of smell avoids the solitary tract and its visceral preoccupation.

16.050 Identify the CORRECT association:
- A. Lateral vestibulospinal tract - excitatory to flexor muscles.
- B. Lateral corticospinal tract - part of the system controlling primarily axial musculature.
- C. Corticopontine fibers - arise mainly from the temporal lobe.
- D. Reticulospinal tracts - principal pathway controlling digital muscles.
- E. Tectospinal tract - controls head-eye reflex movements.

E is the correct answer.
The lateral vestibulospinal tracts are concerned with postural muscle reflex activity in response to vestibular stimuli. They are excitatory to extensor muscles. The lateral corticospinal tracts control the extremities rather than the axial musculature. Corticopontine fibers arise from all lobes of the cerebrum. Reticulospinal tracts have multiple functions such as control of autonomic nuclei, selective gating of sensory input, participation in the control of gamma loop function and muscle tone, etc. However, these fibers are not the principle fibers controlling digital muscles. This is the duty of the lateral corticospinal tracts.

16.051 Select the correct association:
- A. Cingulum - connects Broca's and Wernicke's areas.
- B. Tectospinal tract - projects to all levels of the spinal cord.
- C. Lateral vestibulospinal tract - postural muscle inhibition.
- D. Fornix - projects to the habenula.
- E. Ventral corticospinal tract - mainly controls axial muscles.

E is the correct answer.
See previous answer. Broca and Wernicke are connected by the arcuate (superior longitudinal) fasciculus. The tectospinal tract must move only the head, thus extends only through the cervical cord. The fornix joins hippocampus with hypothalamus and septal region, not the habenula. The main responsibility of the lateral corticospinal tracts is the operation of the extremities and their digits. Axial musculature is innervated by the ventral corticospinal tracts.